CHARACTER ENCYCLOPEDIA

WRITTEN BY SHARI LAST

CONTENTS

SPIDER-MAN

10 Spider-Man (Marvel)
11 Peter Parker
12 Spider-Man (Marvel Cinematic Universe)
13 Other Spider-Men
14 MJ
15 Ned Leeds
16 Vulture
17 The Shocker
18 Mysterio
19 Elemental Illusions
20 Green Goblin (Marvel)
21 Doc Ock (Marvel)
22 Green Goblin (Marvel Cinematic Universe)
23 Doc Ock (Marvel Cinematic Universe)
24 Electro
25 Lizard
26 Spider-Man Variants
28 May Parker
29 J. Jonah Jameson
30 *Bugle* Employees
32 Spider-Man (Miles Morales)
33 Young Spider-Man
34 Spin
35 Ghost-Spider
36 Super Villains
37 Spider-Ham
38 Spider-Man 2099
39 Spider-Man Noir
40 Venom
41 Venomized!
42 Spider-Girl
43 Spider-Woman
44 Scarlet Spider
45 Carnage
46 Sandman
47 Scorpion
48 Comic-Book Foes
50 Gwen Stacy
51 Mary Jane Watson
52 Comic-Book Allies

THE AVENGERS

56 Iron Man
57 Tony Stark
58 Scuba Iron Man
59 Space Iron Man
60 Iron Man's Armor
62 Captain America
63 Steve Rogers
64 Zombie Captain
65 Captain America (Sam Wilson)
66 War Machine
67 Iron Spider
68 Thor
69 The Mighty Thor
70 Hulk
71 Bruce Banner
72 Black Widow
73 Hawkeye
74 Falcon
75 Vision
76 Wanda Maximoff
77 Pietro Maximoff
78 Black Panther (T'Challa)
79 Black Panther (Shuri)
80 Ant-Man
81 The Wasp
82 Doctor Strange
83 Other Stranges
84 Star-Lord
85 Gamora
86 Rocket
87 Groot
88 Drax
89 Mantis
90 Captain Marvel
91 She-Hulk

VILLAINS

94 Obadiah Stane
95 Whiplash
96 Aldrich Killian
97 Trevor Slattery
98 Hydra Agent
99 Ultron
100 Crossbones
101 The Collector
102 Ronan
103 Taserface
104 Ayesha
105 Adam Warlock
106 Loki
107 Hela

108 Grandmaster
109 Gorr
110 Thanos
111 Chitauri and Outriders
112 Ebony Maw
113 Cull Obsidian
114 Proxima Midnight
115 Corvus Glaive
116 Erik Killmonger
117 Ulysses Klaue
118 King Namor
119 Attuma
120 Yellowjacket
121 Ghost
122 Baron Karl Mordo
123 Taskmaster
124 Wenwu
125 Razor Fist
126 Death Dealer
127 Talos
128 Comic-Book Villains

ALLIES

132 Happy Hogan
133 Pepper Potts
134 Helen Cho
135 The Watcher
136 Valkyrie
137 Miek and Korg
138 Winter Soldier
139 Sharon Carter
140 Nick Fury
141 Maria Hill
142 Agent Coulson
143 S.H.I.E.L.D. Agent
144 Wolverine
145 X-Men
146 Yelena Belova
147 Red Guardian
148 Nebula
149 Yondu Udonta
150 Okoye
151 Nakia
152 M'Baku
153 Ironheart
154 Shang-Chi
155 Katy
156 Xialing
157 The Ancient One
158 Wong
159 America Chavez
160 Maria Rambeau
161 Monica Rambeau
162 Ajak
163 Sersi
164 Ikaris
165 Sprite
166 Gilgamesh
167 Thena
168 Phastos
169 Kingo
170 Druig
171 Makkari
172 Index

KEVIN FEIGE

CHAPTER ONE
SPIDER-MAN
Doctor Strange
Ned Leeds
Green Goblin
Spider-Man

A BITE FROM A SPIDER gives Peter Parker the powers he needs to take on New York's most evil villains. But Spider-Man's web-slinging, wisecracking, wall-crawling work doesn't stop there. He teams up with many other heroes to save the country, the planet, the universe—and even the Spider-Verse!

SPIDER-MAN

COMIC-BOOK HERO

SPIDER-MAN'S FEARLESS MINIFIGURE swings into action in dozens of LEGO® sets. There are a lot of Spider-Man minifigures, and most of them wear his iconic red-and-blue Spidey-Suit. This minifigure is based on Spidey's comic-book look, with black webbing details and a small spider emblem on his chest.

Spidey mask with web printing

Printed details on arms

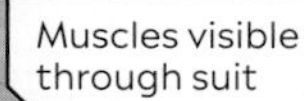

SUPER STATS

LIKES: Stopping villains, being responsible
DISLIKES: Crime
FRIENDS: Iron Man, Ghost-Spider
FOES: Green Goblin, Sandman, Doctor Octopus
SKILLS: Spider-powers, spider-sense, enhanced durability
GEAR: Spidey-Suit, web-shooters

SET: 76172 Spider-Man and Sandman Showdown
YEAR: 2021

STUCK ON YOU
Spider-Man has fought his way out of many sticky situations by trapping villains in his LEGO webs.

DID YOU KNOW?

Spider-Man was the first-ever Super Hero to be licensed and made in LEGO form.

PETER PARKER

FRIENDLY NEIGHBORHOOD SUPER HERO

Hat looks like a Spidey mask that has been pulled up

Hoodie print continues on the back

Belt with Spidey emblem on the buckle

SURPRISE! Under Spider-Man's mask is a teenager named Peter Parker. After being bitten by a radioactive spider, Peter develops super-powers, which he uses for good. His minifigure is wearing Peter's first Spidey-Suit—a homemade outfit of a blue hoodie and red mask.

SUPER STATS

LIKES: Math, rock music
DISLIKES: Bullies
FRIENDS: Mary Jane Watson, Aunt May
FOES: J. Jonah Jameson
SKILLS: Brilliant mind, scientific knowledge, photography
GEAR: Spider Lair, Spider vehicles

SET: 76175 Attack on the Spider Lair
YEAR: 2021

SPIDER HQ

Peter Parker is just like any ordinary kid. But there is one major difference—he has his own Super Hero lair!

SPIDER-MAN

SPIDER AVENGER

SPIDER-MAN'S *NO WAY HOME* minifigure wears a fancy black-and-red suit with integrated nanotech. But when his Spidey-Sense starts tingling, it's not the nanotech he relies on. Instead, Peter's desire to help people is what drives him as he meets five interdimensional villains who are all out to get him. Yikes!

Gold trim shows integrated nanotech

Gold spider symbol spreads across suit

Dual molded printed legs are fan favorites

SUPER STATS

LIKES: Helping people
DISLIKES: Ruining his friends' lives
FRIENDS: MJ, Ned Leeds, Aunt May, Happy Hogan
FOES: Vulture, Mysterio, Thanos, Green Goblin
SKILLS: Spider-sense, spider-powers, enhanced durability
GEAR: Spidey-Suit, web-shooters

SET: 76185 Spider-Man at the Sanctum Workshop
YEAR: 2021

FINDING THE FUNNY
Spidey always makes his friends laugh, even on missions! He enjoys exploring some of the amusing junk stored in the basement of the Sanctum.

OTHER SPIDER-MEN

MULTIVERSE WEB-SLINGERS

AFTER A SPELL GOES WRONG, anyone who knows the identity of Spider-Man gets sucked into Peter Parker's world—including Spider-Men minifigures from other dimensions! Peter is shocked to meet two other versions of himself, who seem very different. But it doesn't take long to see how much they have in common.

FRIENDLY NEIGHBORHOOD SPIDER-MAN
Living up to his title, this Spider-Man is friendly and kind. He teaches Peter to show mercy—even to Green Goblin.

AMAZING SPIDER-MAN
This Spidey admits that he's never fought an alien, though he wishes he had! He encourages Peter to keep going, no matter how bad things seem.

Unique head piece has large eyes

Long-legged spider emblem

Webbing details in titanium

Dual molded blue-and-red legs

MJ

SPIDER-MAN'S BEST FRIEND

IT DOESN'T TAKE LONG for super-smart MJ to figure out that her classmate (who mysteriously disappears whenever a villain shows up) is none other than Spider-Man! MJ's minifigure is fully supportive of Peter's Super Hero-ing, and she bravely fights alongside him in three LEGO sets.

Head piece used on all three MJ minifigures

Face printed with inquisitive expression

Plain and comfortable sweater

SUPER STATS

LIKES: Reading, churros
DISLIKES: Swinging over the city, making friends
FRIENDS: Peter Parker, Ned Leeds
FOES: Vulture, Lizard
SKILLS: Highly observant, investigative mind
GEAR: None

SET: 76261 Spider-Man Final Battle
YEAR: 2023

SAVED BY SPIDEY
Being friends with a Super Hero can place you in danger. Luckily, MJ is saved by Spidey—though he is from an alternate reality!

NED LEEDS

SPIDER-MAN'S OTHER BEST FRIEND

Long, floppy hair frames Ned's face

SPIDER-MAN CAN ALWAYS count on his best friend Ned, whether they're building LEGO models together or stopping villains. Ned's minifigure wears everyday clothes, but he often performs the job of a hero—from being Spidey's "guy in the chair" to opening interdimensional portals.

DID YOU KNOW?

Ned's confidence has grown over the years. His first minifigure had one happy and one worried face, while this minifigure has two smiling expressions.

Letterman jacket with buttons and stripe details

Hands sometimes spark magic

SUPER STATS

LIKES: Building with LEGO bricks, gummy worms
DISLIKES: The paparazzi
FRIENDS: Peter Parker, MJ
FOES: Vulture, Lizard
SKILLS: Computer hacking, magic
GEAR: Laptop, sling ring (borrowed)

SET: 76261 Spider-Man Final Battle
YEAR: 2023

STRANGE MAGIC

Ned is astonished to discover he possesses some "Doctor Strange magic." His minifigure is even rescued by Strange's cloak during the battle on the Statue of Liberty!

VULTURE

FLYING FOE

CRIMINAL ADRIAN TOOMES builds weapons using alien Chitauri technology and sells them on the illegal market. His Vulture minifigure is always scheming, but somehow, Spider-Man is always there to stop him. It's very frustrating!

Head piece printed with black flight mask and glowing green eyes

Fur-lined flight jacket

Breathing apparatus attached under head piece

Jetpack harness continues around back of torso

SUPER STATS

LIKES: Making money
DISLIKES: Lazy people
FRIENDS: Shocker, Tinkerer
FOES: Spider-Man
SKILLS: Strong business sense, flying expertise
GEAR: Exo-suit, motorized wings, Chitauri weapons

SET: 76195 Spider-Man's Drone Duel
YEAR: 2021

DID YOU KNOW?

There are five Vulture minifigures and they each come with different wings. Two have a single LEGO wing element (in different colors), while three come with large, buildable wings.

WINGED VILLAIN
Vulture has his very own piece of alien tech—an exo-suit with motorized wings that give him a soaring advantage in battle.

THE SHOCKER

ELECTRIFYING ENEMY

HERMAN SCHULTZ IS known as the Shocker, thanks to his Chitauri gauntlet, which fires blasts of electricity. His minifigure appears in only one set, but that gives him more than enough time for a supercharged battle against Spider-Man!

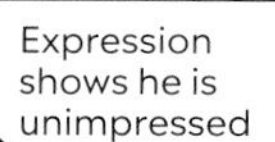

Expression shows he is unimpressed

Crosshatch details on arms

Black jacket with hood printed on the back

SUPER STATS

LIKES: Making money
DISLIKES: Homemade Super Hero suits
FRIENDS: Vulture
FOES: Spider-Man
SKILLS: Top thief, combat skills
GEAR: Shocker Gauntlet

SET: 76083 Beware the Vulture
YEAR: 2017

GETAWAY DRIVER
Driving a van loaded with dangerous weapons, the Shocker struggles to make a clean escape when Spider-Man shows up!

MYSTERIO

WANNABE SUPER HERO

QUENTIN BECK SEEMS like a nice guy, but don't be fooled! A special effects expert, he creates the illusion that Earth is under attack. Then his minifigure swoops to the rescue as Mysterio, a friendly interdimensional hero. But Spider-Man is under no illusion: Mysterio must be stopped.

Light-blue round helmet unique to Mysterio

Elaborate Super Hero suit is meant to look otherworldly

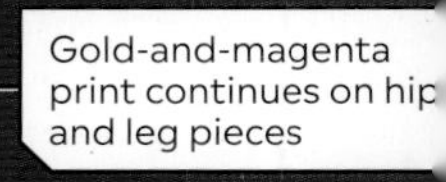

Protective mesh on suit

SUPER STATS

LIKES: Being famous
DISLIKES: Not being appreciated by Tony Stark
FRIENDS: None
FOES: Spider-Man, Tony Stark
SKILLS: Special effects, engineering, genius mind
GEAR: Drones and illusion-tech

SET: 76184 Spider-Man vs. Mysterio's Drone Attack
YEAR: 2021

DRONE ATTACK

Mysterio uses Stark technology and thousands of drones to cast his illusions. Spidey finally defeats him during a huge battle on London's Tower Bridge.

DID YOU KNOW?

Mysterio wanted his suit to look heroic, so it was modeled on the costumes of Thor, Doctor Strange, and Black Panther.

ELEMENTAL ILLUSIONS

HYDRO-MAN AND MOLTEN MAN

BUILDABLE BADDIE
Fiery Molten Man's LEGO form is a buildable figure rather than a minifigure. He appears in the 2019 LEGO set Molten Man Battle (set 76128).

DID YOU KNOW?
Hydro-Man's minifigure comes with matching LEGO bricks that can create a towering wave effect for him to stand on.

WHILE HYDRO-MAN AND MOLTEN MAN are illusions cast by the villain Mysterio, their LEGO forms are perfectly real! Wreaking havoc around the world, Hydro-Man whips up a swirling whirlpool in Venice and Molten Man causes a fiery scene in Prague. Both times, Spider-Man swings to the rescue!

GREEN GOBLIN

SPIDER-MAN'S ARCHENEMY

AN EXPERIMENT GONE WRONG turned scientist Norman Osborn into a seriously unhinged villain. When his Green Goblin alter ego takes over, his minifigure dons a frightful mask and costume and causes mayhem for the people of New York. Time and again, Spider-Man is the only one brave enough to stand up to him!

Goblin ears attached to hat piece

Don't confuse this grin for friendliness

Clothing torn in battle

Legs printed with armored kneepads

SUPER STATS

LIKES: Being powerful, crime
DISLIKES: Anyone who tries to stop him
FRIENDS: None
FOES: Spider-Man
SKILLS: Cunning mind, super-strength, agility
GEAR: Goblin Glider, Green Goblin armor, pumpkin bombs

SET: 76175 Attack on the Spider Lair
YEAR: 2021

GOLDEN-EYED TREASURE

The very first Green Goblin minifigure was released in 2003, with an all-green suit and gold eyes. For 10 years, it was the only LEGO Green Goblin available.

DOC OCK

YOUNG OCTOPUS

YOUNG DOC OCK FEATURES in the animated series Marvel's *Spidey and His Amazing Friends*. Her minifigure might have shorter legs and longer hair than the traditional Marvel Comics villain, but her green suit, long tentacles, and mischievous grin are enough to show that this Doc Ock is up to no good!

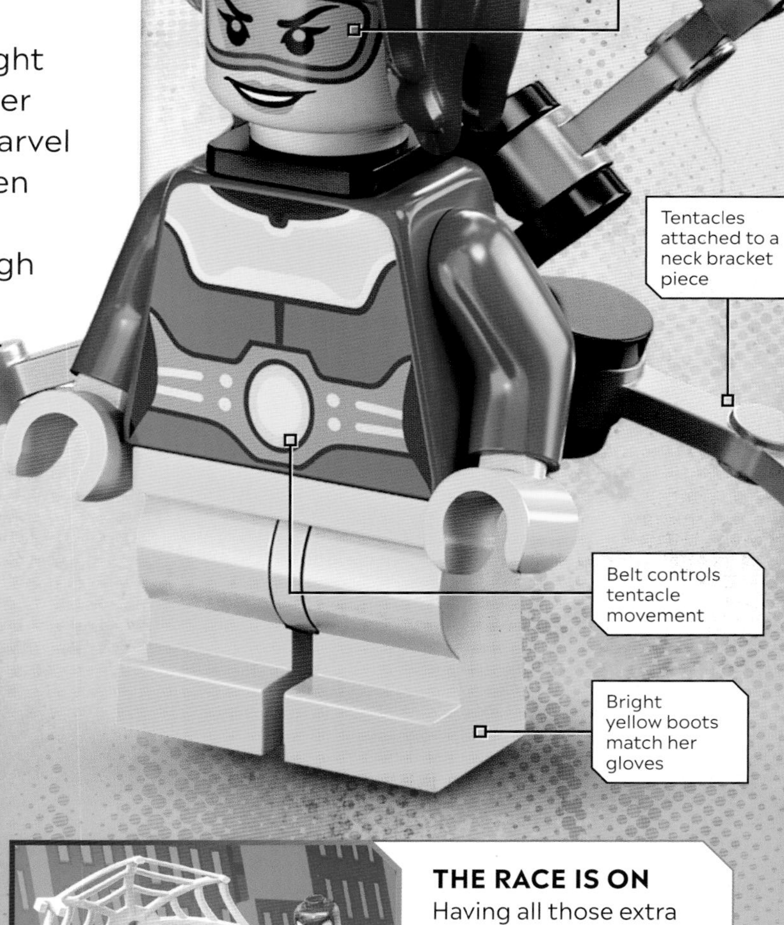

SUPER STATS

LIKES: Stealing stuff
DISLIKES: Getting caught!
FRIENDS: Octobots
FOES: Spidey, Spin, Ghost-Spider
SKILLS: Plotting and scheming
GEAR: Mechanical tentacles, Octobots

SET: 10789 Spider-Man's Car and Doc Ock
YEAR: 2023

THE RACE IS ON
Having all those extra arms is useful during a diamond heist. If only Doc Ock had a few extra legs to help her outrun Spidey!

GREEN GOBLIN

VILE VILLAIN

GOBLIN GEAR

Green Goblin zooms into battle on his Goblin Glider. He's known for throwing pumpkin bombs—which are not as amusing as they sound!

Ripped remnants of a purple hoodie

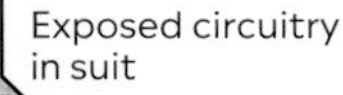

Belt with storage pouch

SUPER STATS

LIKES: Causing chaos
DISLIKES: Weakness
FRIENDS: None
FOES: Spider-Man
SKILLS: Genius mind, superhuman strength, speed, healing
GEAR: Goblin Glider, Green Goblin armor, pumpkin bombs

SET: 76261 Spider-Man Final Battle
YEAR: 2023

DID YOU KNOW?

There are only two glider-less Green Goblin sets: one shows the moment Osborn became the Green Goblin and the other comes with a battle mech instead.

THE GREEN GOBLIN'S TERRIFYING minifigure has been Spider-Man's archenemy for years, but this is his first—and only—MCU minifigure. He still wears the green and purple colors of the Goblin, but this time Norman Osborn's human face is visible.

DOC OCK

TENTACLED TERROR

HOLDING ON

Doc Ock uses his 20-piece LEGO tentacles to climb buildings and grab enemies during battle. They attach to his body with a bracket around his neck.

SUPER STATS

LIKES: Complaining, science
DISLIKES: Removing his glasses
FRIENDS: Spider-Man
FOES: Spider-Man
SKILLS: Genius mind
GEAR: Mechanical tentacles

SET: 76261 Spider-Man Final Battle
YEAR: 2023

A SCIENTIST POSSESSED by his own mechanical tentacles, Otto Octavius is bitter and cruel. Until, that is, he meets a new Spider-Man in a different reality, who figures out how to cure him! Back in control of his own mind, Doc Ock uses his tentacles to help Spidey cure even more of his foes.

ELECTRO

POWER-HUNGRY FOE

TRANSPORTED ACROSS THE Multiverse, Electro's minifigure is excited by the power he can sense in the air. Charging his body with sizzling energy, he refuses Spider-Man's potential cure and heads out across the city to absorb all the delicious electricity it has to offer.

Electricity crackles across face

Arc Reactor enhances Electro's powers

Tactical vest features buckles and cables

SUPER STATS

LIKES: Power!
DISLIKES: Eels
FRIENDS: Doctor Octopus, Sandman
FOES: Spider-Man
SKILLS: Controlling electricity, flight
GEAR: Arc Reactor (temporarily)

SET: 76261 Spider-Man Final Battle
YEAR: 2023

OVERLOAD
Electro can't resist the power of Spider-Man's Arc Reactor. When he integrates it into his suit, the results are truly shocking!

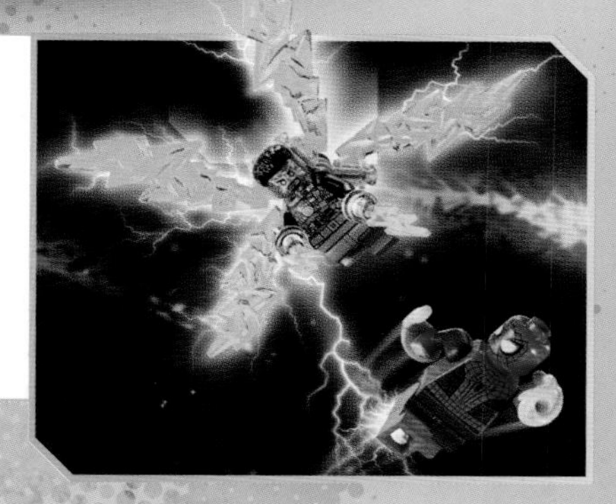

LIZARD

REPTILIAN FOE

DID YOU KNOW?

While this is Lizard's first minifigure, his first-ever LEGO appearance was back in 2013 as a big fig character in the LEGO *Marvel Super Heroes* video game.

DR. CURT CONNORS MERGED reptile DNA with his own during an unsuccessful experiment. A frequent enemy of Spider-Man, Lizard makes his first LEGO appearance in 2024, where his scaly minifigure battles Spider-Man at the Statue of Liberty.

Right hand was damaged in battle

Tail piece fits over hip piece

Reptilian scales cover body

Sharp claws printed on feet

SUPER STATS

LIKES: Turning people into lizards
DISLIKES: When people are surprised he can speak!
FRIENDS: Sandman, Doctor Octopus
FOES: Spider-Man, Ned Leeds, MJ
SKILLS: Reptilian abilities, high intelligence
GEAR: None

SET: 76280 Spider-Man vs. Sandman: Final Battle
YEAR: 2024

FINAL BATTLE
During his big battle with Spidey, Lizard opens his mouth and takes a snappy bite... into a cure! Finally, Connors returns to his original human body.

SPIDER-MAN

VARIANTS GALLERY

SPIDER-MAN HAS SWUNG into action in more than 100 LEGO sets, and his minifigures rock a lot of different looks. Here are some of the coolest, most iconic, and rarest of Peter Parker's minifigures.

FIRST EVER

The very first Spider-Man minifigure, this was released in 2002 in three LEGO sets.

Silver webbing printed on hip piece

DID YOU KNOW?

Spider-Man is one of the two Super Hero minifigures most often featured as a Comic-Con exclusive minifigure.

BALACLAVA BOY

This rare minifigure shows Peter in his first homemade Spidey-Suit.

Balaclava hides most of Peter's face

Faded spider design on sweater

MOST COMMON

A very classic minifigure, this Spidey has appeared in the most sets: 14!

Small spider emblem

Plain blue legs

RAREST
This 2013 Comic-Con exclusive is considered the rarest Spider-Man variant of all time.
Halftone pattern in eyes
Very detailed torso print
Suit print continues on legs
MIGHTY MICRO
This Mighty Micro figure has short legs and a cheeky, winking face!
Bold lines are common on Mighty Micro minifigures
Large red spider emblem printed on back of torso

STEALTH SUIT
Wearing his black-and-green Stealth Suit from the comics, this minifigure is often called "Big Time Spider-Man" by fans.
PS4 EXCLUSIVE
Another Comic-Con exclusive, this minifigure takes the form of the Spidey character from the PlayStation 4 game.
Silver shading in eyes
Every piece of this minifigure is unique
Rare white spider emblem
Printed details on hips, legs, and feet
Green lights glow when suit is in Camo Mode
Suit has the ability to absorb sound
No webbing details on suit

MAY PARKER

AUNT AND ALLY

THERE MAY BE INFINITE MAYS in The Multiverse, but there are only four May minifigures—and they all show her as she appears in the classic comic books. No matter how many heroes Spider-Man meets, May will always be the person he admires most in the world!

Warm and kind face

Sweater also worn by Hermione Granger from LEGO® Harry Potter™

Practical dark blue pants

DID YOU KNOW?

This May minifigure is visiting the *Daily Bugle* building in set 76178. She carries a plate of wheat cakes—probably freshly baked for her favorite nephew!

SUPER STATS

LIKES: Baking
DISLIKES: Shirking responsibility
FRIENDS: Peter Parker, Otto Octavius
FOES: Doctor Octopus, Green Goblin
SKILLS: Strong-willed, kind, compassionate
GEAR: None

SET: 76178 *Daily Bugle*
YEAR: 2021

NOT AGAIN!
Being related to a Super Hero, May often finds herself in danger, which is why her alternate face print shows a scared expression!

J. JONAH JAMESON

EDITOR-IN-CHIEF

Hair is always sticking up

DID YOU KNOW?

This minifigure's appearance is based on Jameson's first minifigure, released in 2004.

J. JONAH JAMESON IS PROUD to be the Editor-in-Chief of *The Daily Bugle* newspaper. His minifigure wears a crisp white shirt and black suit, signaling how seriously he takes his job. He oversees the newspaper's reporters and photographers, but there is one question he obsesses over day and night: Who is Spider-Man?

Alternate face shows a glob of spiderwebs covering his mouth

Vest with blue buttons and trim

SUPER STATS

LIKES: Boxing, tai chi
DISLIKES: Spider-Man, Spider-Man, and Spider-Man
FRIENDS: *The Daily Bugle* employees
FOES: Spider-Man
SKILLS: Investigating, media expertise
GEAR: Microphone, computer

SET: 76178 *Daily Bugle*
YEAR: 2021

EDITOR AT WORK
Jameson's office is where he displays his editing awards for all to see. His desk, however, is cluttered with his Spidey research.

BUGLE EMPLOYEES

NEWSPAPER GALLERY

***DAILY BUGLE* EMPLOYEES** are always busy, searching for the latest scoop and prepping for their news bulletins. But none of them realize that the truth behind their paper's biggest mystery—who is Spider-Man?—can be found in a small office down the hall!

DID YOU KNOW?

These *Daily Bugle* employees are so dedicated to their jobs that they appear in only one LEGO set—*Daily Bugle* (set 76178).

ROBBIE ROBERTSON
This high-ranking editor is friends with the boss, J. Jonah Jameson, but secretly likes Spider-Man.

BETTY BRANT
A former secretary to J. Jonah Jameson, Betty is now a TV reporter.

Never seen without his gold-rimmed glasses

BEN URICH

An investigative reporter, Ben is good at uncovering secret identities... most of the time.

Checkered jacket worn at the office

Head piece also used for MJ in set 76129

AMBER GRANT

As a freelance photographer, Amber often crosses paths with Peter Parker.

Satchel holds her camera

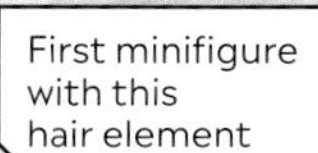

First minifigure with this hair element

Diamond pattern continues on the back

RON BARNEY

Ron is a junior journalist who once wrote an article about the Silver Surfer.

A DAY AT THE OFFICE

Television screens and framed newspaper printouts adorn the walls of the cluttered *Daily Bugle* newsroom.

SPIDER-MAN

MILES MORALES

SPIDEY SIMILARITIES
Like Peter Parker, Miles makes his first Spidey-Suit himself. He chooses a red-and-black color scheme, using a hoodie to cover his head.

SUPER STATS

LIKES: Art, graffiti
DISLIKES: Snobby schools
FRIENDS: Spider-Man, Aaron Davis, Spider-Gwen
FOES: Kingpin, the Spot
SKILLS: Spider-sense, spider-powers
GEAR: Spidey-Suit

SET: 76244 Miles Morales vs. Morbius
YEAR: 2023

DID YOU KNOW?

This minifigure's masked head piece is shared by two other Miles Morales/Spider-Man minifigures—both of which wear a red hoodie on top.

ANOTHER REALITY, ANOTHER TEENAGER, and another radioactive spider bite! This is Miles Morales, and his Spider-Man minifigure protects the streets of Brooklyn in its red-and-black Spidey-Suit.

YOUNG SPIDER-MAN

BOY HERO

Big eyes with bold black outline

Bold spider emblem on simple web-lines

YOUNG SPIDEY APPEARS in Marvel's *Spidey and His Amazing Friends* animated series. His minifigure wears the iconic Spidey-Suit, though it features simplified details and shorter legs—ideal for younger LEGO fans. Spidey joins his two best friends, Spin and Ghost-Spider, to stop their not-so-friendly foes.

SUPER STATS

LIKES: Basketball, insects
DISLIKES: Too many amusement park rides
FRIENDS: Spin, Ghost-Spider
FOES: Doc Ock, Rhino, Green Goblin
SKILLS: Spider-sense, web-slinging, wall-crawling
GEAR: Spidey-Suit

SET: 10782 Hulk vs. Rhino Truck Showdown
YEAR: 2022

SPIDER CAR
Though he's just a kid, Spidey drives a cool car to catch crooks. It even has its own web-shooters!

SPIN

SUPER KID

MILES MORALES goes by the Super Hero nickname Spin in the animated TV series Marvel's *Spidey and His Amazing Friends*. His minifigure has short legs, large eyes, and strikingly bold red details on its spider-suit.

DID YOU KNOW?

There are three Spin minifigures from the TV series, and they each feature a different color spider emblem on their chests.

Back of torso is printed with a red spider

Red webbing shoots from wrists

SUPER STATS

LIKES: Painting, pistachio ice cream
DISLIKES: Forgetting to eat breakfast
FRIENDS: Spidey, Ghost-Spider
FOES: Doc Ock, Green Goblin
SKILLS: Spider-sense, cloaking power, arachno-sting
GEAR: Spider-suit, webs

SET: 10781 Spider-Man's Techno Trike
YEAR: 2022

SPIDER FRIENDS
Spin enjoys painting in his spare time. His latest artwork shows him and his two besties: Spidey and Ghost-Spider.

GHOST-SPIDER

GWEN STACY

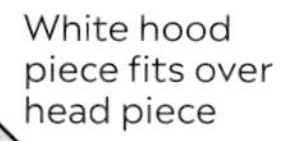

TAKE TO THE SKIES!
In set 10784, Ghost-Spider likes to hang out in the Spidey Webquarters, where she keeps her speedy Ghost-Copter.

SUPER STATS

LIKES: Playing the drums, animals
DISLIKES: Unsolved mysteries
FRIENDS: Spidey, Spin
FOES: Doc Ock, Rhino
SKILLS: Spider-sense, gliding abilities
GEAR: Spider-suit, glider

SET: 10783 Spider-Man at Doc Ock's Lab
YEAR: 2022

GWEN STACY IS ALSO KNOWN AS Ghost-Spider. Her minifigure joins Spidey and Spin on their adventures, dressed in her purple, pink, and white spider-suit. Her white hood and head piece gives her a particularly ghostly look!

SUPER VILLAINS

YOUNG FOES

YOUNG SPIDEY and his friends might be kids, but they take on all sorts of villains, from giggling goblins to moody robots.

SANDMAN
True to his name, Sandman transforms himself into sand in his endless efforts to capture Spidey and his friends at the beach.

Hat with goblin ears attached

GREEN GOBLIN
Young Green Goblin, also known as Gobby, has shorter legs than the other Green Goblin minifigures, but he wears the same nasty grin on his mischievous adventures!

Belt with two large pockets

"G" buckle on belt

Body can turn to sand particles

RHINO
Look out! Rhino's minifigure uses all his strength to crack eggs and hurl them at Spidey and the other heroes in set 10791.

Rhino horns on helmet

Special LEGO piece created just for Zola

Green face is on a screen in middle of body

Tough, bulky shoulder guards

ZOLA
This robot villain attacks Spidey's HQ in a large yellow mech. He brings along a tiny Zola clone, too!

SPIDER-HAM

PIG HERO

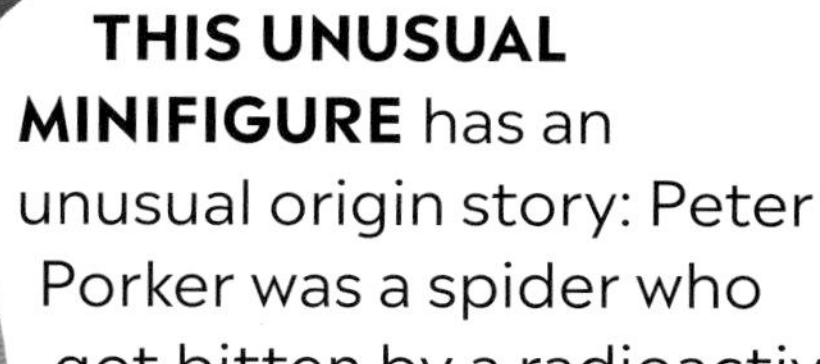

THIS UNUSUAL MINIFIGURE has an unusual origin story: Peter Porker was a spider who got bitten by a radioactive pig! Known to the world as Spider-Ham, he fights for justice alongside various Spider-Men from across the Spider-Verse.

Snout features spider-eye nostrils

Pig head mold printed with webbing

Classic spider-suit similar to Peter Parker's

DID YOU KNOW?

The pig-shaped head piece was created for Spider-Ham and has been used for just one other minifigure: Pork Grind, a Venomized version of Spider-Ham.

SUPER STATS

LIKES: Hot dogs
DISLIKES: When people assume that he can't talk
FRIENDS: Spider-Man, Captain America
FOES: Venom, Ducktor Doom
SKILLS: Spider-powers
GEAR: Spider-suit, web-spinner gauntlets

SET: 76151 Venomosaurus Ambush
YEAR: 2020

SNACK TIME
Even during battle, Spider-Ham always makes time for a tasty snack—and to crack plenty of silly jokes!

SPIDER-MAN 2099

HERO FROM THE FUTURE

SPIDER-MAN 2099 IS NOT ONLY from an alternate reality—he is also from the future! Genius geneticist Miguel O'Hara wears his futuristic spider-suit in just one set, where his minifigure joins Peter Parker in battle against their common enemies.

Huge red spider emblem dominates costume

Comic-book halftone pattern printed on torso and head

Suit made from Unstable Molecule fabric

SUPER STATS

LIKES: Experimenting with AI
DISLIKES: People who abuse their power
FRIENDS: Spider-Man
FOES: Sandman, Vulture, the Spot
SKILLS: Spider-powers, enhanced vision and hearing, can produce webs
GEAR: Spider-suit, web-shooters

SET: 76114 Spider-Man's Spider Crawler
YEAR: 2019

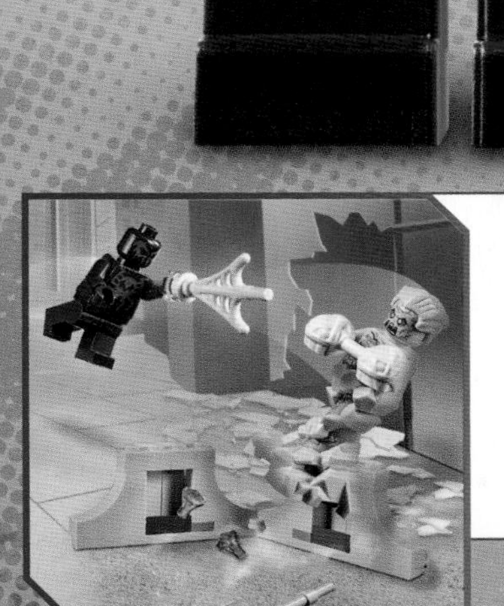

SANDMAN GETS STUCK
Shooting webs at enemies comes naturally to Spider-Man 2099 because his body produces them without any need for artificial web fluid. Watch out, Sandman!

SPIDER-MAN NOIR

HERO FROM THE PAST

ANOTHER TIME-TRAVELING SUPER HERO, this version of Spider-Man is from the 1930s, when everything was in black and white. His monochrome minifigure wears cool spider shades and a trenchcoat. His somber appearance reflects his serious personality.

DID YOU KNOW?

The hat Noir wears is a classic LEGO piece that is most often worn by cowboy and explorer minifigures.

Shirt is old-fashioned in style

Untucked shirt detail on hip piece

Leg piece also used for Nick Fury minifigure

SUPER STATS

LIKES: Solving crimes
DISLIKES: Colorful things
FRIENDS: Spider-Man
FOES: Venom, Norman Osborn
SKILLS: Spider-powers, detective instinct
GEAR: None

SET: 76150 Spiderjet vs. Venom Mech
YEAR: 2020

PILOT SKILLS
Despite being from the past, Spider-Man Noir flies a Spiderjet. He uses his amazing spider-sense to pilot the craft.

VENOM

EVIL ALIEN

THE ALIEN VENOM is one of Spider-Man's most slippery foes. This is mostly because its natural form is a pile of black goo, but also because it can change shape and bond with living beings. Venom first bonded with Spidey, which is why its minifigures all feature spider imagery.

DID YOU KNOW?

The earliest Venom minifigures grin with their teeth together. But more recent Venom variants have a more open mouth, revealing a bright tongue.

SUPER STATS

LIKES: Bonding with powerful beings
DISLIKES: Complicated feelings
FRIENDS: Whoever it's bonded with right now
FOES: Spider-Man, Iron Man
SKILLS: Shape-shifting, healing
GEAR: None

SET: 242104 Venom
YEAR: 2021

LOOKS LIKE VENOM

Venom is known for its evil grin, long red tongue, and black tentacles. Many of its LEGO forms feature these characteristics, including this monstrous mech!

VENOMIZED!

CORRUPTED CHARACTERS

VENOM TAKES ON the appearance of whomever it bonds with, boosting their powers while slowly turning them evil. Venomized minifigures are easy to spot—they keep some of their original form but also show telltale signs that they have been corrupted.

PORK GRIND

The pigheaded Pork Grind is a Venomized version of Spider-Ham. The villain often wields a huge mallet and hates spinach.

No mouth printed on head piece

Torso is used for other Venom minifigures

VENOM SPIDEY

This 2012 Comic-Con exclusive shows Spider-Man in his Venom suit. It took Peter Parker a while to realize the suit was having a bad effect on him!

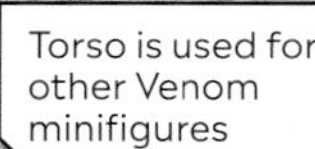

GOBLIN VENOM

Villains get Venomized, too! Venomized Goblin is partly transformed, making him more terrifying than ever. Look out!

Half-Venom, half-Iron Man helmet printing

Iron Man's armor transforms into a Venom costume

Venom emblem spreads right across chest

Goblin's regular green legs are fully Venomized

IRON VENOM

As Venom corrupts Iron Man, the armored Avenger's helmet and suit take on Venom's black-and-white color scheme.

SPIDER-GIRL

SUPER SPIDER

HIGH-SCHOOL STUDENT Anya Corazón uses her spider-powers to help people. She has also gone by the name Araña and sometimes wears a red-and-white suit. Her minifigure teams up with Spider-Man to chase Doctor Octopus in just one set, making her a rare member of the LEGO Spider family.

Unique head piece printed with black mask

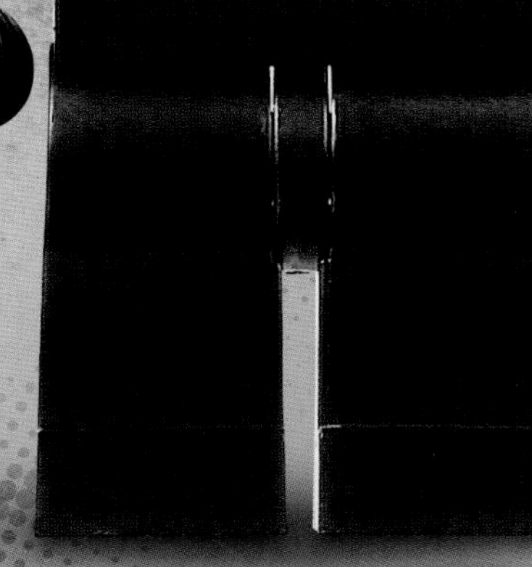

Large spider emblem wraps around to the back of the torso

Webbing produced in her forearm

SUPER STATS

LIKES: Computers, rap music
DISLIKES: Bullies, evil scientists
FRIENDS: Spider-Man
FOES: Doctor Octopus
SKILLS: Spider-powers, camouflage abilities
GEAR: Spider-suit

SET: 76148 Spider-Man vs. Doc Ock
YEAR: 2020

WEB WARRIOR
From her walking spider, Spider-Girl shoots webs at Doc Ock. She hopes to trap his tentacles in a tangle!

DID YOU KNOW?

In a different reality, there's another Spider-Girl named May Parker, who is Peter Parker's daughter!

SPIDER-WOMAN

RARE HERO

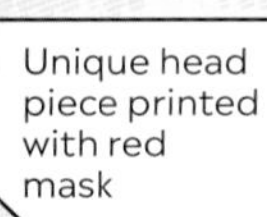

Hands can shoot "venom blasts" of electricity

Torso printing continues on the hip piece

Dual molded legs show Spider-Woman's yellow boots

ULTIMATE VERSION
An alternate Spider-Woman swings into action in set 76057. Her red-and-white minifigure is based on Marvel's Ultimate Comics.

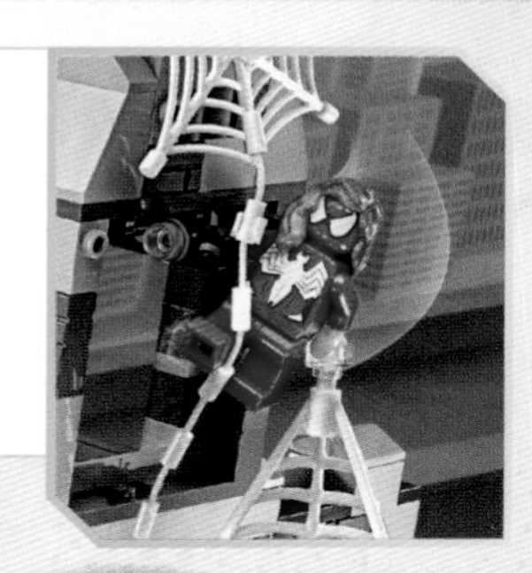

DID YOU KNOW?

While both minifigures on this page are versions of Spider-Woman, the red-and-white minifigure is referred to as Spider-Girl in set 76057.

SUPER STATS

LIKES: Butter
DISLIKES: Rats
FRIENDS: Doctor Strange
FOES: Morgan le Fay
SKILLS: Spider-powers, resistant to poison
GEAR: Spider-suit

SET: Comcon027 Spider-Woman
YEAR: 2013

JESSICA DREW, AKA SPIDER-WOMAN, makes a very rare appearance in her red-and-yellow comic-book suit. Only 350 copies of this Spider-Woman minifigure were released, as a giveaway at the 2013 San Diego Comic-Con, making it one of the rarest LEGO Super Hero minifigures.

SCARLET SPIDER

SPIDER-MAN CLONE

BRIDGE RESCUE
Scarlet Spider uses his sticky skills to help rescue Aunt May from Scorpion on a crumbling bridge.

Same head piece as Spider-Woman, who appears in the same set

Wristbands can store web pellets

Hood element is a separate piece

SUPER STATS

LIKES: Having blond hair (under his mask)
DISLIKES: Clone confusion
FRIENDS: Spider-Man, Aunt May
FOES: Scorpion, Green Goblin
SKILLS: Spider-powers, exceptional balance
GEAR: Web-shooters, web pellets

SET: 76057 Spider-Man: Web Warriors Ultimate Bridge Battle
YEAR: 2016

A CLONE OF SPIDER-MAN, Ben Reilly was created by a villain. Fortunately, Reilly decides to become a good guy—and he takes the name Scarlet Spider. His minifigure wears a red suit under a hooded vest, complete with spider emblem and utility belt. He joins Spider-Man in one LEGO set.

CARNAGE

SAVAGE ALIEN

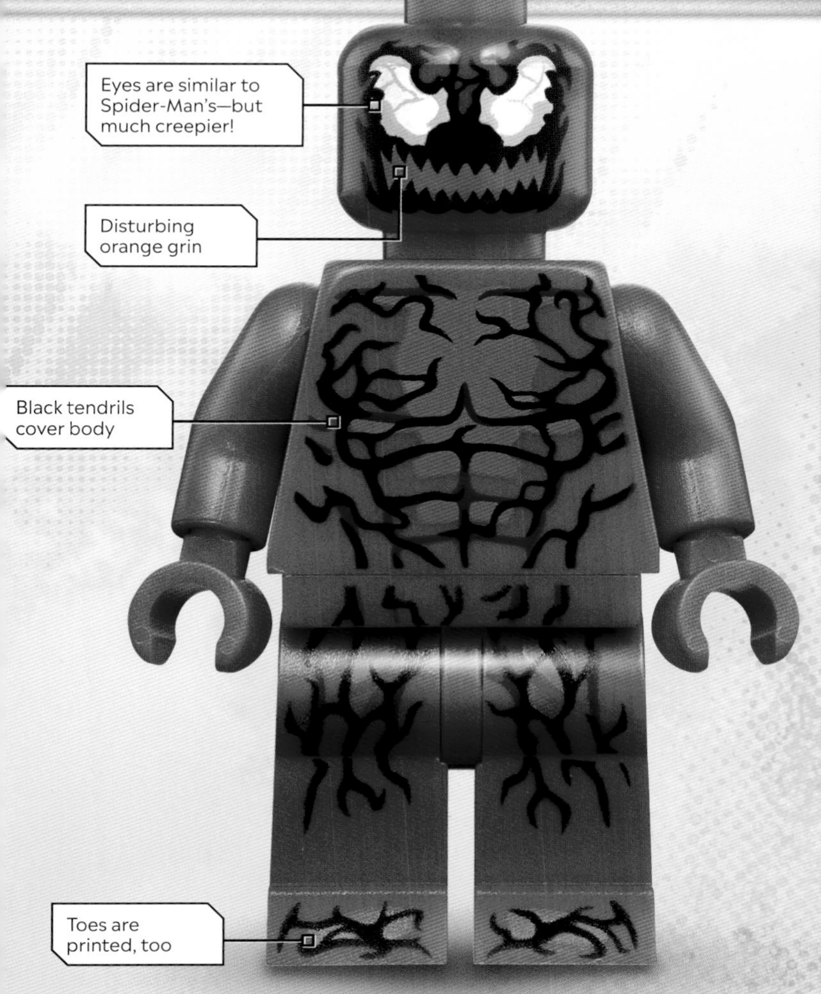

Eyes are similar to Spider-Man's—but much creepier!

Disturbing orange grin

Black tendrils cover body

Toes are printed, too

SUPER STATS

LIKES: The color red
DISLIKES: Family
FRIENDS: None
FOES: Spider-Man, Venom
SKILLS: Shape-shifting
GEAR: None

SET: 76173 Spider-Man and Ghost Rider vs. Carnage
YEAR: 2021

POWER DOWN
When Carnage wraps his tentacles around a power generator, Spider-Man prepares for a shocking battle!

LIKE VENOM, Carnage is an alien that bonds with human hosts. Also like Venom, he finds Spider-Man working hard to thwart his every move. Carnage can transform parts of himself into weapons, but his minifigure prefers to sprout bright-red tentacles!

SANDMAN

MAN OF SAND

A RADIATION ACCIDENT leaves criminal William Baker with special super-powers. He can transform his body into sand, in all shapes and sizes! This Sandman minifigure doesn't even come with leg pieces because he is partly transformed. Instead, he stands on a swirling mass of sand!

DID YOU KNOW?

All five of Sandman's minifigures wear his comic-book-inspired striped sweatshirt—though each one is partly or fully turned into sand!

SUPER STATS

LIKES: The beach, sandcastles, football
DISLIKES: Bullies
FRIENDS: Sinister Six
FOES: Spider-Man
SKILLS: Transforming his body into sand, turning his arms into sand-weapons
GEAR: None

SET: 76172 Spider-Man and Sandman Showdown
YEAR: 2021

SAND-FORMATION!
In set 76114, a unique version of Sandman shows him completely turned to sand. His skin has lost all color and his mouth is obscured by dripping grains of sand!

SCORPION

STINGING SUPER VILLAIN

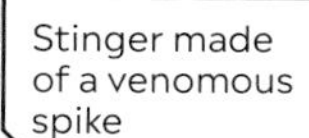

AFTER UNDERGOING AN EXPERIMENT, private investigator Mac Gargan gains super-powers—but also loses his grip on reality. His minifigure wears a scorpion suit, complete with a unique tail piece, to attack his foes. Watch out for that stinger!

SUPER STATS

LIKES: Being paid a lot of money
DISLIKES: Time travel
FRIENDS: Justin Hammer, Norman Osborn
FOES: Spider-Man, J. Jonah Jameson
SKILLS: Super-strength, wall-crawling
GEAR: Scorpion suit, motorized tail

SET: 76057 Spider-Man: Web Warriors Ultimate Bridge Battle
YEAR: 2016

MICRO TAIL
A Mighty Micros version of Scorpion uses the same LEGO tail piece. However, the tip of this tail is yellow, not gray.

COMIC-BOOK FOES

ROGUES' GALLERY

SPIDER-MAN BATTLES super villains of all kinds, from masked sorcerers to highly competitive hunters. These villains have a range of super-powers, a range of goals, and a range of incredible costumes!

Homemade armored suit is super-strong

Wrist-mounted weapons built into the suit

BEETLE

Beetle joins Doctor Doom for an attack on *The Daily Bugle* HQ. He uses his winged suit to engage Spidey in an aerial battle!

KRAVEN THE HUNTER

To prove he is the best hunter of all time, Kraven decides to catch someone that no one else can—Spider-Man! But Spidey continues to be the one that got away...

Iron mask bonded to Doom's face

DOCTOR DOOM

Sorcerer and tech genius Doctor Doom makes his only LEGO appearance, so far, when he attacks Spider-Man at *The Daily Bugle* offices in set 76005.

Fur collar also worn by Korg minifigure

Animal print on hips and legs

Green tunic continues on leg pieces

DID YOU KNOW?

Several minifigures wear the same tattered cape as Hobgoblin, including Corvus Glaive. But Hobgoblin is the only minifigure to wear it in bright orange.

HOBGOBLIN

Inspired by the Green Goblin, Hobgoblin's minifigure wears a unique orange version of the Goblin suit as he embarks on a life of crime.

RHINO

Rhino's minifigure comes with a huge, strong Rhino Mech in set 76037. But under his Rhino armor, he just looks like an ordinary man in a vest!

Muscles visible beneath vest

Harness to attach heavy rhino suit

MORBIUS

Dr. Michael Morbius might have batlike powers, but he's not a total villain. His minifigure sometimes fights alongside Spidey.

GWEN STACY

COMIC-BOOK GIRLFRIEND

AT UNIVERSITY, GWEN STACY meets and falls in love with fellow science major Peter Parker. Gwen doesn't know Peter's spider-shaped secret, so she can't understand why her minifigure so often finds herself in danger!

Freckles printed on head piece

Light, casual jacket keeps Gwen warm

SUPER STATS

LIKES: Helping the police solve cases
DISLIKES: Spider-Man
FRIENDS: Peter Parker, Harry Osborn
FOES: Green Goblin
SKILLS: Scientific mind, kindness
GEAR: None

SET: 76178 *Daily Bugle*
YEAR: 2021

FAMILY FRIEND

In set 76059, Spider-Man tries to helps Gwen's dad, Captain George Stacy, too!

DID YOU KNOW?

In a different Spider-Verse reality, another Gwen Stacy is bitten by a spider and becomes Spider-Woman—also known as Spider-Gwen (or Ghost-Spider)!

MARY JANE WATSON

OTHER COMIC-BOOK GIRLFRIEND

Head piece unique to this minifigure

MARY JANE HAS KNOWN Peter Parker for years, and they eventually fall in love. Her minifigure is shocked to discover her boyfriend is Spider-Man, but she quickly proves herself to be a resourceful, supportive, and brave ally!

Spidey heart tank top shows how Mary Jane feels about her favorite Super Hero!

SUPER STATS

LIKES: Acting
DISLIKES: Being described only as "nice"
FRIENDS: Peter Parker
FOES: Green Goblin
SKILLS: Keeping secrets, business sense
GEAR: None

SET: 76016 Spider-Helicopter Rescue
YEAR: 2014

DID YOU KNOW?

There are four other Mary Jane minifigures. They are all based on the character as she appears in the *Spider-Man* trilogy movies.

SCARY FOR MARY
It's easy to see why Mary Jane would need her alternate face. It shows her sheer terror when she's captured by the Green Goblin!

COMIC-BOOK ALLIES

HEROES' GALLERY

SPIDER-MAN HAS MANY friends and allies who are always willing to team up and take on a super villain or two. These cool, colorful minifigures even have unique super-powers of their own!

No other minifigure wears this hair piece in the same color

Unique head piece printed with red mask

FIRESTAR
With flaming red hair and a bright orange suit, Firestar can control heat in all its forms.

IRON FIST
Gaining powers from a dragon's heart, Iron Fist channels his inner energies into enhanced abilities, including superhuman strength!

Dragon emblem on chest

Fist glows with inner energy

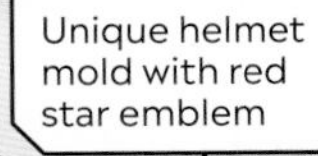

NOVA
A young Super Hero with cosmic powers, Nova's minifigure wears the uniform of the Nova Corps intergalactic police force.

Three disks show rank of Centurion

GHOST RIDER

There are three Ghost Rider minifigures. This one shows Johnny Blaze, who made a deal with scary Mephisto. That's why he looks this way!

BLACK CAT

Most of the time, Black Cat is a stealthy cat burglar. But her minifigure joins Spidey in one set to help him capture some villains.

POWER MAN

Luke Cage, aka Power Man, is known for working together with Iron Fist. In set 76016, however, he teams up with Spidey.

WHITE TIGER

Using her super-strength, reflexes, and heightened senses, White Tiger joins Spidey in a battle against Doc Ock and Vulture.

CHAPTER TWO
THE AVENGERS

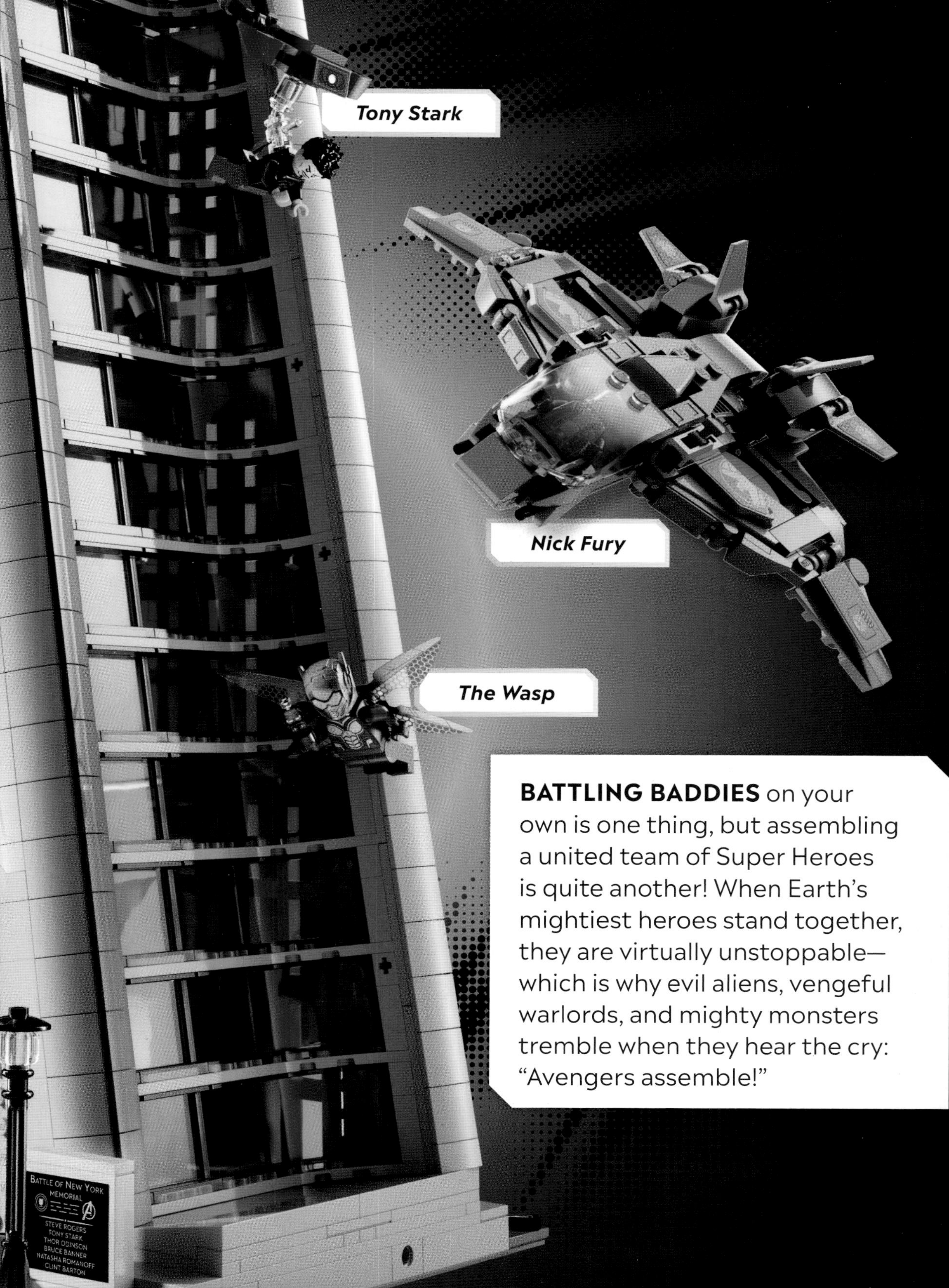

BATTLING BADDIES on your own is one thing, but assembling a united team of Super Heroes is quite another! When Earth's mightiest heroes stand together, they are virtually unstoppable—which is why evil aliens, vengeful warlords, and mighty monsters tremble when they hear the cry: "Avengers assemble!"

IRON MAN

ARMORED AVENGER

MIDMISSION MECHANIC

Iron Man's genius mind and tech knowledge enable him to repair vehicles, even in the middle of a dangerous mission!

SUPER STATS

LIKES: High-tech gadgets
DISLIKES: Being called "Metal Man"
FRIENDS: Avengers, Spider-Man
FOES: Thanos
SKILLS: Tech knowledge, strength
GEAR: Armored suit, many gadgets

SET: 76248 The Avengers Quinjet
YEAR: 2023

DID YOU KNOW?

Iron Man's Mark 7 armor has been recreated in LEGO form three times: in 2012 (the first-ever Iron Man minifigure), in 2016, and in this 2023 version.

STANDING STRONG in his iconic red-and-gold armor, Iron Man's minifigure defends Earth in dozens of LEGO® sets. Powered by the Arc Reactor embedded in his chest, Iron Man's armor has undergone dozens of upgrades over the years, as has his minifigure.

TONY STARK

GENIUS INVENTOR

DID YOU KNOW?
Out of 50 Iron Man minifigures, only a handful don't wear full Iron Man armor. It's clear where Stark's priorities lie!

CEO OF STARK INDUSTRIES, Tony Stark is known for being rich, brilliant, and outspoken. He might have created more than 90 Iron Man armored suits, but his minifigures sometimes appear wearing casual, everyday clothing or—in this case—a heavy metal sweatshirt!

Expression is never fully relaxed

Sweater design includes a LEGO helmet with visor

Comfortable dark gray pants

SUPER STATS

LIKES: Inventing new things, taking risks
DISLIKES: Being told what not to do
FRIENDS: Pepper Potts, Happy Hogan, Nick Fury
FOES: Obadiah Stane, Aldrich Killian
SKILLS: Genius mind
GEAR: Stark Industries' lab and resources

SET: 76194 Tony Stark's Sakaarian Iron Man
YEAR: 2021

OFF DUTY
It's rare to see Tony Stark relax. Even when he's not wearing his Iron Man armor, he's busy tweaking the various pieces of technology in his lab.

SCUBA IRON MAN

UNDERWATER AVENGER

IRON MAN IS KNOWN for having many LEGO suits, but one of the most unusual is a scuba suit! Scuba Iron Man battles evil in just one set from 2016, making it a pretty rare minifigure. The sand-green underwater suit features an armored chestplate with Arc Reactor, and unique front, back, and leg printing.

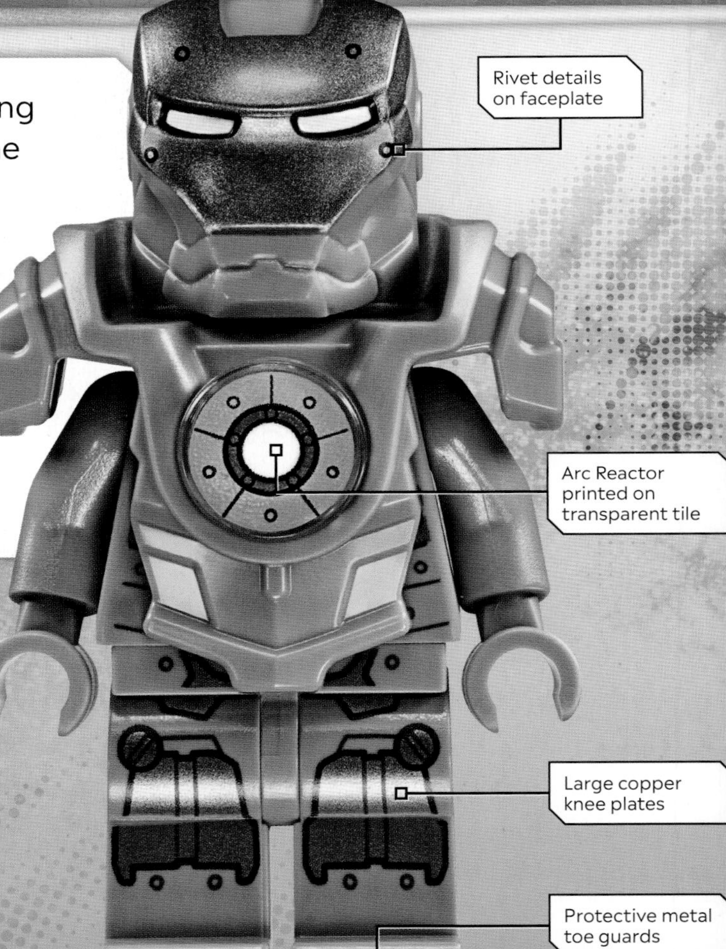

SUPER STATS

LIKES: Swimming
DISLIKES: Hydra
FRIENDS: Captain America
FOES: Red Skull
SKILLS: Strategic mind
GEAR: Scuba suit

SET: 76048 Iron Skull Sub Attack
YEAR: 2016

DARING DIVER

Iron Man's scuba suit and fire-blasting weapons prove incredibly useful when he chases a Hydra diver beneath the waves.

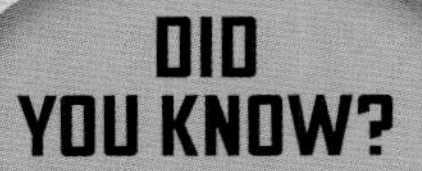

DID YOU KNOW?

This minifigure has nine parts (some are on the back). Most exciting for collectors is the exclusive 1x1 Arc Reactor chest piece.

SPACE IRON MAN

ZERO GRAVITY HERO

Blue trim around eyes adds comic-book effect

Gold-plated armor visible beneath outer space suit

Armor panels act as protection from the vacuum of space

FORTUNATELY FOR IRON MAN, Tony Stark's tech works perfectly well in space. In this set, Iron Man wears a cool white suit and unique helmet for deep space combat. With metallic gold and silver elements, rear air filters, and foot thrusters, this minifigure is out of this world!

SUPER STATS

LIKES: Space travel
DISLIKES: Motion sickness
FRIENDS: Captain America
FOES: Hydra
SKILLS: Space flight, advanced technology
GEAR: Space suit with propulsion capabilities

SET: 76049 Avenjet Space Mission
YEAR: 2016

SPACE FACE-OFF

Space Iron Man launches into battle with Space Thanos. However, the biggest test is whether Iron Man's thrusters will be a match for Thanos's stud shooters.

IRON MAN'S ARMOR

SUIT GALLERY

TONY STARK'S MIND is always working. His never-ending ideas inspire him to create more than 90 different armors! Some are specialized for space or underwater combat (like those on the previous pages), while others are upgraded with new gadgets or futuristic tech. His minifigures wear different suits, too. Those on this page are based on his comic-book exploits. Those opposite are based on movie appearances.

DID YOU KNOW?

Toy Fair Iron Man is the only minifigure to have the Iron Man helmet printed on the head piece.

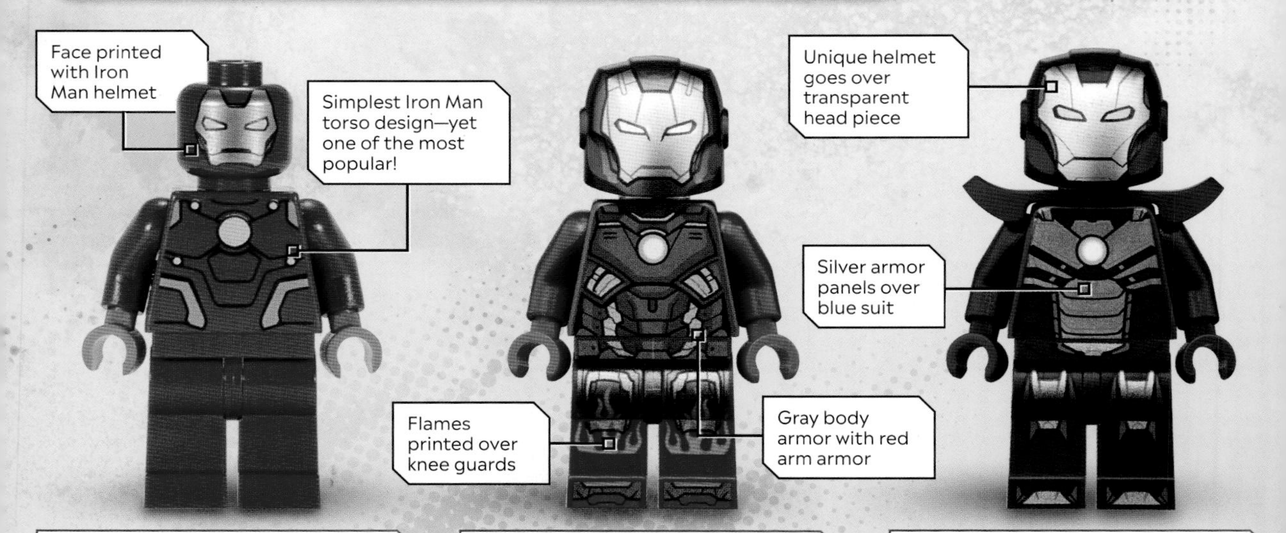

TOY FAIR IRON MAN

Released as a Toy Fair exclusive in 2012, this is the first Iron Man minifigure ever, and also the rarest. Only 125 were made!

BLAZER

This unique armor is known as Blazer due to the flame design on the legs. The helmet, head, and torso are all unique.

TAZER

The only blue Iron Man minifigure suit is known as the Tazer. It is also the only suit with this style of shoulder armor.

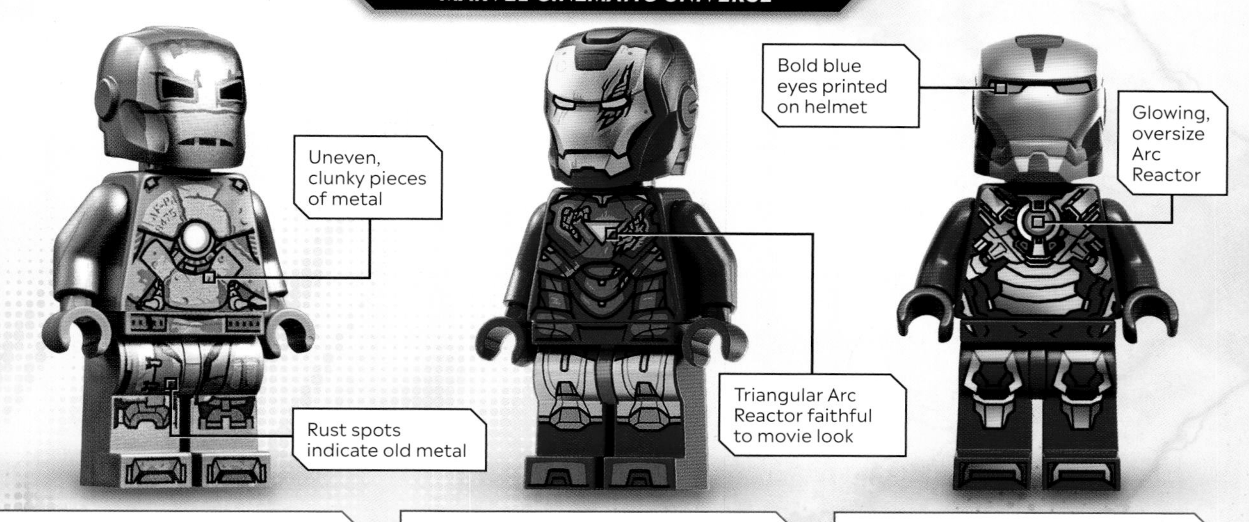

MARK 1
Stark built his first-ever Iron Man suit from scrap metal. This minifigure is printed to show all of the scratches and dents on the armor.

MARK 6
This armor first appeared in Marvel Studios' *Iron Man 2*, after Stark created a new element for his more powerful Arc Reactor.

MARK 17
Iron Man's "Heartbreaker" armor is a fan favorite because of its detailed printing. Stark wears it in Marvel Studios' *Iron Man 3*.

MARK 25
Code-named Striker, this armor is designed for construction. Its arms are very powerful—good for flinging villains away!

MARK 33
Another fan favorite, the Silver Centurion minifigure is made up of 12 separate elements. These include a unique head piece showing Stark with a bruised face.

MARK 46
Stark uses nanotechnology in this suit, making it smarter. He wears it in Marvel Studios' *Captain America: Civil War*, where the collapsible helmet comes in handy.

CAPTAIN AMERICA

STEVE ROGERS—A NATION'S HERO

A SYMBOL OF HOPE for his country—and the world—Captain America stands for justice and honor. His minifigure wears the stars and stripes with pride as one of the leading members of the Avengers. Cap's super-strength is legendary, as are his super-strong opinions.

2023 head piece includes blue helmet printing

Stars and stripes motif on costume

Red gloves fit with the American flag color scheme

Vibranium shield is Cap's only weapon

SUPER STATS

LIKES: Justice
DISLIKES: Dishonesty
FRIENDS: Black Widow, Falcon
FOES: Thanos, Hydra
SKILLS: Strength, speed, healing powers
GEAR: Vibranium shield

SET: 76248 The Avengers Quinjet
YEAR: 2023

DID YOU KNOW?

Out of 22 Cap minifigures, only one wears something other than red, white, and blue: the *Endgame* version wears a white Team Suit instead.

CIVIL WAR
The Avengers don't always assemble on the same side. There was a time when Cap went head-to-head with Iron Man over whether Super Heroes should make their identities public.

STEVE ROGERS

ALTERNATE UNIVERSE HERO

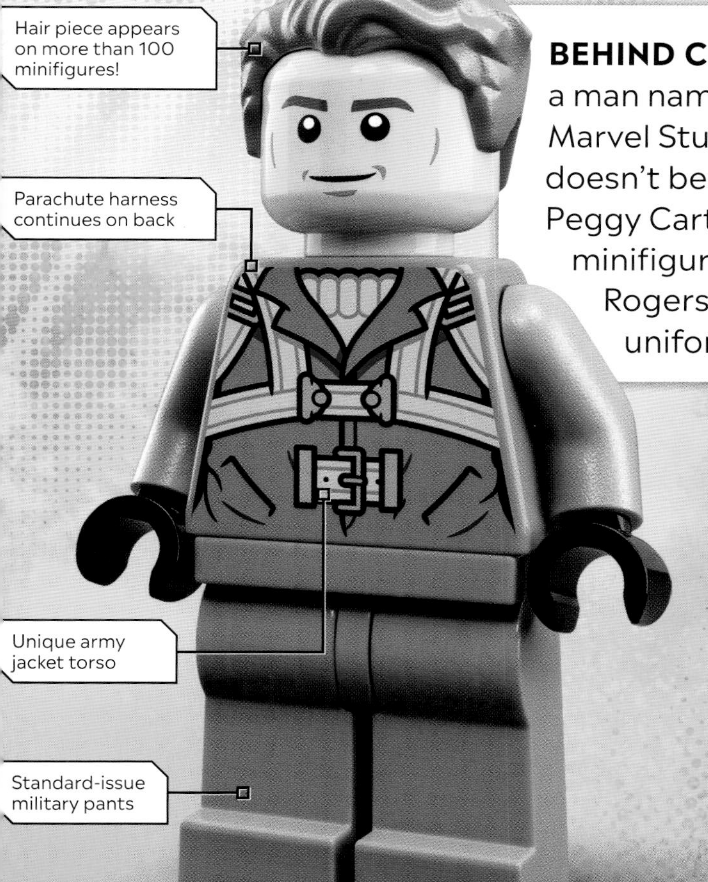

Hair piece appears on more than 100 minifigures!

Parachute harness continues on back

Unique army jacket torso

Standard-issue military pants

BEHIND CAPTAIN AMERICA'S mask is a man named Steve Rogers. But in the Marvel Studios' *What If...?* series, Rogers doesn't become the first Avenger at all—Peggy Carter does instead! This *What If...?* minifigure is the only one to show Steve Rogers in his standard WWII soldier uniform.

SUPER STATS

LIKES: Music, patriotism
DISLIKES: Bullies
FRIENDS: Bucky Barnes, Peggy Carter
FOES: Red Skull
SKILLS: Combat, strength
GEAR: WWII weapons

SET: 76201 Captain Carter & the Hydra Stomper
YEAR: 2021

ALTERNATE CAP

In set 76201, ordinary soldier Steve Rogers joins forces with the nation's favorite Super Hero, Captain Carter. Together, they battle the evil Red Skull and secure the powerful Tesseract.

DID YOU KNOW?

Steve Rogers might not be Cap in this alternate reality, but his heroism and courage still help save the day!

ZOMBIE CAPTAIN

ALTERNATE CAP

AFTER BEING BITTEN by a zombie in the Marvel Studios' *What If...?* series, Captain America is transformed. His yellow eyes gleam with rage as he sets out to attack his fellow Super Heroes. There have been many Cap minifigures, but none as terrifying as this one!

SUPER STATS

LIKES: Being evil
DISLIKES: Trains
FRIENDS: None
FOES: Sharon Carter, Bucky Barnes
SKILLS: Zombie rage
GEAR: Shield (though it is used against him!)

SET: 71031-13 LEGO Minifigures—Marvel Studios Series
YEAR: 2021

ZOMBIE HUNTER
Watch out, Cap! Zombie Hunter Spidey comes in the same minifigure collection—and he's wearing Doctor Strange's Cloak of Levitation.

CAPTAIN AMERICA

A NEW CAP

Red goggles similar to Falcon minifigure's

Unique suit printed with patriotic details

Famous shield now wielded by Wilson

SAM WILSON USED TO BE known as Falcon, but his minifigure recently got a super new look—as Captain America! Taking over as Cap after Steve Rogers retires, Wilson takes his new job very seriously. His latest minifigure wears a unique version of the Captain America suit, carries the iconic shield, and was created specially for this book!

SUPER STATS

LIKES: Running in the park
DISLIKES: Too much chit-chat in battle
FRIENDS: Steve Rogers, Black Widow, Winter Soldier
FOES: Hydra, Thanos, Karli Morgenthau
SKILLS: Combat training
GEAR: EXO-7 Falcon wingsuit, vibranium shield

SET: Exclusive minifigure in DK's LEGO *Marvel Character Encyclopedia*
YEAR: 2024

FLIGHT READY
Wilson brings his own skills to his new role as America's favorite hero. He uses his Falcon wingsuit—now in patriotic red, white, and blue—to soar into battle for his country.

WAR MACHINE

ARMORED WARRIOR

JAMES "RHODEY" RHODES is an expert pilot and good friend of Tony Stark—but he is also a Super Hero known as War Machine! His minifigure used to wear an old Iron Man suit but now stands strong in high-tech armor of his own.

LEGO helmet opens, like Iron Man's

Arc Reactor powers suit

Suit arms loaded with automatic weapons

Reinforced supportive leg armor

SUPER STATS

LIKES: A proper chain of command
DISLIKES: When Tony Stark hacks into his computer
FRIENDS: Tony Stark, the Avengers
FOES: Obadiah Stane, Thanos
SKILLS: Immense strength, combat experience, piloting expertise
GEAR: War Machine armor, Stark Sonic Cannon, Ex-Wife Missile

SET: 76216 Iron Man Armory
YEAR: 2022

SHOULDER CANNON
Each of War Machine's minifigures is armed with a shoulder cannon. Most fire LEGO stud pieces.
This minifigure, from set 242213, has two shoulder-mounted weapons!

IRON SPIDER

HIGH-TECH HERO

Four long pincers can be activated by AI

Gold trim around spider symbol

Bulletproof nanotech fabric withstands the pressures of space

THERE ARE MORE THAN 20 different Spider-Man minifigures, but none are as multifunctional as Iron Spider! Swinging into his new role as an Avenger, this Spidey wears an armored nanosuit gifted to him by his mentor, Tony Stark. Using AI, Spidey can control the suit with just his mind.

SUPER STATS

LIKES: Nanotechnology
DISLIKES: Being sent home during battle
FRIENDS: The Avengers
FOES: Thanos, Doc Ock
SKILLS: Spider skills
GEAR: Iron Spider suit, with parachute and missiles

SET: 76108 Sanctum Sanctorum Showdown
YEAR: 2018

SPIDER SKILLS

Spider-Man uses all of his leaping, diving, and swinging skills to help the Avengers as they try to stop Thanos from collecting the Infinity Stones.

THOR

GOD OF THUNDER

BEING AN IMMORTAL Super Hero is tiring work. The God of Thunder and former king of Asgard is also an Avenger and Guardian of the Galaxy. It's no wonder Thor was thinking of retiring! Thor's minifigure, however, refuses to rest when innocent people are in danger, and Thor dons his armor and red cape once more.

Gray hairs in beard

Scale armor beneath leather suit

Royal Asgardian disks on chestplate

Long, flowing red cloak

SUPER STATS

LIKES: Being worthy
DISLIKES: Sibling rivalry
FRIENDS: The Mighty Thor, Avengers, Guardians of the Galaxy
FOES: Hela, Thanos
SKILLS: Strength, durability, combat
GEAR: Stormbreaker axe, Mjolnir hammer (formerly)

SET: 76209 Thor's Hammer
YEAR: 2022

HAMMER IT HOME
Thor was almost as crushed as his Mjolnir hammer after it was destroyed. But his new weapon, the Stormbreaker axe, harnesses lightning with even more power!

THE MIGHTY THOR

WIELDER OF MJOLNIR

Dual molded silver helmet with blonde hair

Intricate armor of Asgardian design

DOCTOR JANE FOSTER steps up to defend New Asgard—an act of bravery that proves her worthy of wielding Thor's legendary hammer, Mjolnir. Her mighty minifigure rocks the "Thor" look, though The Mighty Thor's armor and winged helmet are unique to her.

Silver armor details printed on reddish-brown hips and legs

SUPER STATS

LIKES: Having a good Super Hero catchphrase
DISLIKES: Dinner dates
FRIENDS: Thor
FOES: Gorr
SKILLS: Strength, speed, sharp reflexes, flight
GEAR: Mjolnir

SET: 76207 Attack on New Asgard
YEAR: 2022

BATTLING GORR
The Mighty Thor takes on the vengeful Gorr at the Gates of Eternity. Gorr taunts the hero, calling her "Lady Thor." In response, she smashes his Necrosword to bits.

HULK

ANGRY AVENGER

BIG, GREEN, AND SOMETIMES MEAN, Hulk transforms whenever scientist Bruce Banner gets angry. Hulk's muscular LEGO big fig towers over regular minifigures, which is often scary enough to make them turn and flee. Fortunately for the Avengers, Hulk is on their side!

Muscles are part of the torso mold

Fists pack a powerful punch!

SUPER STATS

LIKES: Smashing things
DISLIKES: Restrictive clothing
FRIENDS: The Avengers
FOES: Loki, Thanos
SKILLS: Immense strength and durability
GEAR: None needed

SET: 76131 Avengers Compound Battle
YEAR: 2019

HULK SMASH!
Hulk performs his legendary Hulk Smash move against Thanos while defending Avengers Compound. That's gonna hurt!

BRUCE BANNER

BRILLIANT SCIENTIST

Unique face print shows Banner's serious expression

ALMOST THE COMPLETE opposite of the Hulk, Bruce Banner is a soft-spoken genius. His minifigure would much rather dive into his lab research than leap onto the battlefield—though, to his dismay, that choice isn't always under his control.

DID YOU KNOW?

There are two Bruce Banner minifigures, both featuring an alternate face with an angry expression and ominous green eyes.

Jacket worn over checkered shirt

SUPER STATS

LIKES: Classical music, science
DISLIKES: Getting angry
FRIENDS: Natasha Romanoff, Tony Stark
FOES: Loki
SKILLS: Genius intellect
GEAR: None

SET: 76247 The Hulkbuster: The Battle of Wakanda
YEAR: 2023

HULKBUSTER

United by their genius minds, Bruce Banner and Tony Stark become friends. Stark eventually builds the Hulkbuster suit, which helps Banner control his rage.

BLACK WIDOW

SUPER SPY

TRAINED AS AN ASSASSIN, Natasha Romanoff is a founding member of the Avengers. Her minifigure never hesitates to leap into action, using her martial arts training to defeat Earth's enemies in more than 15 LEGO sets.

DID YOU KNOW?

Black Widow's minifigures always wear combat suits, apart from one. In the 2023 LEGO Marvel Avengers Advent Calendar, she wears a woolly Christmas sweater!

Torso and leg design are exclusive to this minifigure

Glowing blue piping on suit

Suit from Marvel Studios' *Avengers: Age of Ultron* movie

SUPER STATS

LIKES: Honesty
DISLIKES: Being underestimated
FRIENDS: Bruce Banner, Hawkeye, the Avengers
FOES: Thanos, Taskmaster, Ultron
SKILLS: Combat, espionage, weapons specialist
GEAR: Twin batons

SET: 76260 Black Widow & Captain America Motorcycles
YEAR: 2023

NEED FOR SPEED
One of Black Widow's many skills is being an ace motorcycle driver, as proven during a speedy chase after Ultron's soldiers.

HAWKEYE

MASTER MARKSMAN

SPY TURNED AVENGER Clint Barton is always right on target. His skill and precision with a bow and arrow have earned him the code name Hawkeye. His minifigure carries a quiverful of arrows, which range from putty projectiles to high-tech rocket drones. Ready, aim, fire!

S.H.I.E.L.D. emblem on combat vest

Red glove protects the hand that pulls the bowstring

Pockets and straps printed on battlesuit

SUPER STATS

LIKES: Keeping his work and family life separate
DISLIKES: Robot armies
FRIENDS: Black Widow, The Scarlet Witch, the Avengers
FOES: Loki, Ultron, Thanos
SKILLS: Expert marksman
GEAR: Bow and arrows

SET: 76269 Avengers Tower
YEAR: 2023

MOTORCYCLE MAYHEM
Hawkeye is best known for his marksmanship. However, he takes it to a whole new level in set 76067, when he takes aim from a speeding motorcycle!

FALCON

WINGED AVENGER

ARMY VETERAN SAM WILSON meets Steve Rogers running in a park and soon becomes a key member of the Avengers! His minifigure wears an armored wingsuit from Wilson's days as a soldier, which allows him to fly with precision.

Wings slot onto neck

Wings are one solid piece

DID YOU KNOW?

Although Falcon joins the Avengers in several LEGO sets, it is as a completely different minifigure each time!

SUPER STATS

LIKES: Fighting for his country
DISLIKES: Holding a grudge
FRIENDS: Steve Rogers, the Avengers
FOES: Hydra, Crossbones, Thanos
SKILLS: Combat training
GEAR: EXO-7 Falcon wingsuit, red goggles

SET: 76050 Crossbones' Hazard Heist
YEAR: 2016

FLYING HIGH
Falcon's wings give him an undeniable advantage in battle. They even come with a detachable drone on the back, which Falcon names Redwing.

VISION

ANDROID AVENGER

Mind Stone is what powers Vision

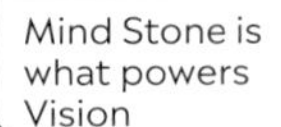

Cape and clasp printed on torso

WITH A BODY MADE FROM vibranium and human cells, and the Mind Stone embedded in his forehead, Vision is a Super Hero like no other! His unique minifigure stands out too, with transparent legs, vibrant, detailed printing, and a transparent flexi-plastic cape. What a vision!

Suit can transform at Vision's will

Unique cape has transparent lower section

SUPER STATS

LIKES: Being polite, logic
DISLIKES: Destruction
FRIENDS: The Scarlet Witch, the Avengers
FOES: Ultron, Thanos
SKILLS: Flight, transformation of matter
GEAR: Mind Stone

SET: 76269 Avengers Tower
YEAR: 2023

SOMETHING ON HIS MIND

When Thanos is hunting for Infinity Stones, Vision heads to Wakanda. Here, Shuri takes on the tricky task of trying to remove the Mind Stone without destroying Vision's mind.

WANDA MAXIMOFF

THE SCARLET WITCH

WANDAVISION

Wanda and Vision build a happy life in the series Marvel Studios' *WandaVision*, but all is not as it seems. It looks like Wanda has been changing her own reality this time!

Alternate face has glowing red eyes

Necklace with black stone

Red leather jacket is worn by all Wanda minifigures

Skirt cloth piece created especially for this minifigure

SUPER STATS

LIKES: Her family
DISLIKES: Making mistakes
FRIENDS: Vision, Hawkeye, Pietro Maximoff
FOES: Doctor Strange, Thanos, Ultron
SKILLS: Altering reality, mind control, energy blasts
GEAR: None

SET: 76266 Endgame Final Battle
YEAR: 2023

WANDA MAXIMOFF IS SO POWERFUL she can change reality. This was great for the Avengers—when Wanda was on their side. But after a terrible tragedy, everything changed. Wanda is now known as The Scarlet Witch, and her minifigure wreaks havoc across The Multiverse!

PIETRO MAXIMOFF

QUICKSILVER

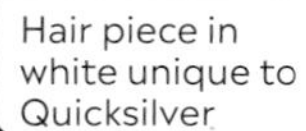

Unique head piece printed with stubble

Muscles visible through thin fabric

Durable, flexible running pants

PIETRO MAXIMOFF RACES into the lives of the Avengers so fast he's just a blur. He uses his super-speed to protect the citizens of Sokovia and his twin sister, Wanda. Pietro appears in just one Marvel movie. Similarly, his minifigure comes in just one LEGO set.

DID YOU KNOW?

Pietro and Wanda got their super-powers by participating in experiments carried out by Hydra's Baron Strucker.

SUPER STATS

LIKES: Defending his country
DISLIKES: War
FRIENDS: Wanda Maximoff
FOES: Ultron
SKILLS: Super-speed
GEAR: None

SET: 76041 The Hydra Fortress Smash
YEAR: 2015

HATER TO HERO

Before teaming up with the Avengers to defeat Ultron, Pietro was not a fan of the group. In fact, he attacked them in a snowy forest.

BLACK PANTHER

FORMER KING OF WAKANDA

IN HIS STATE-OF-THE-ART Black Panther suit, Black Panther stands ready to defend the people of Wakanda—and Earth. Joining forces with the Avengers, he offers his kingdom's technology and resources to help save lives.

Head piece unique to this minifigure

Claw necklace generates suit and mask

Purple glow as the suit absorbs energy

Suit made from vibranium-weave fabric

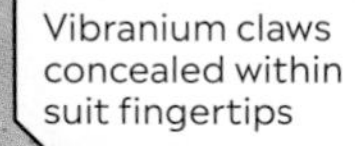

Vibranium claws concealed within suit fingertips

SUPER STATS

LIKES: Protecting his kingdom
DISLIKES: Revenge
FRIENDS: Shuri, Nakia, the Avengers
FOES: Erik Killmonger, Thanos
SKILLS: Enhanced senses, speed, and strength
GEAR: Black Panther suit, vibranium weapons

SET: 76186 Black Panther Dragon Flyer
YEAR: 2021

FEARLESS LEADER

Black Panther leads by example. He is the first to charge into battle in set 76103, when the army of Wakanda defends against Thanos's invading forces.

DID YOU KNOW?

The face of T'Challa, the man behind the Black Panther mask, appears on a *What If...?* minifigure, which shows T'Challa as Star-Lord!

BLACK PANTHER

RULER OF WAKANDA

DID YOU KNOW?

Shuri's Black Panther suit comes with an alternate head piece in set 76214, which shows Shuri's face. Her trademark hair piece is included, too.

SCIENTIST AND INVENTOR

Shuri is the new Black Panther following the death of her brother, T'Challa. Her minifigure wears an all-new Black Panther suit as she sets out to stop a dangerous underwater foe.

SUPER STATS

LIKES: American culture
DISLIKES: Uncomfortable, stiff clothing
FRIENDS: T'Challa, Okoye
FOES: Namor, Attuma
SKILLS: Knowledge of technology, enhanced senses, speed, and strength
GEAR: Vibranium gauntlets

SET: 76214 Black Panther: War on the Water
YEAR: 2022

WARRIOR PRINCESS

Shuri has always fought for justice as a scientist, inventor, and warrior. Before she donned the Black Panther suit, her minifigure charged into battle in traditional Wakandan clothes.

ANT-MAN

SIZE-CHANGING HERO

ANT-MAN'S SPECIAL SUIT enables him to grow big or shrink to the size of an ant. At first, Scott Lang uses his new powers to steal things, but he later puts them to much better use defeating villains. Often, his minifigure is so small that enemies don't even see him coming!

High-tech suit protects against the dangerous effects of changing size

Buttons on gloves activate suit's shrinking function

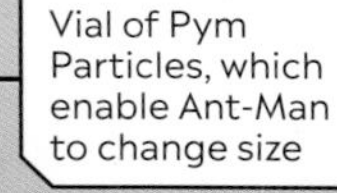

Vial of Pym Particles, which enable Ant-Man to change size

SUPER STATS

LIKES: Magic tricks, karaoke
DISLIKES: Rats, suit malfunctions
FRIENDS: The Wasp, the Avengers
FOES: Yellowjacket, Ghost, Thanos, Kang
SKILLS: Size manipulation, stealth, tactical mind
GEAR: Ant-Man suit, Pym Particles

SET: 76109 Quantum Realm Explorers
YEAR: 2018

QUANTUM EXPLORER

Ant-Man shrinks smaller than anyone has ever done before, eventually ending up in the Quantum Realm. He explores this dangerous place in an insectoid quantum vehicle.

THE WASP

TINY FLYING HERO

HOPE VAN DYNE IS the daughter of Hank Pym, creator of Pym Particles and the Ant-Man suit. Hope teaches Scott Lang to use his shrinking powers, and she dons a suit of her own as The Wasp! Just like Ant-Man, she can shrink to a tiny size. But—as her minifigure loves reminding Ant-Man—she can also fly, thanks to her suit's high-tech wings.

SUPER STATS

LIKES: Making fun of Ant-Man
DISLIKES: Not being invited to battles
FRIENDS: Ant-Man, Hank Pym
FOES: Yellowjacket, Ghost, Kang
SKILLS: Size manipulation, flight
GEAR: Wasp suit

SET: 76269 Avengers Tower
YEAR: 2023

CHANGING SIZE

Both Ant-Man and The Wasp come in different LEGO forms. Here, Scott turns into a giant buildable figure while The Wasp's microfigure flutters by his side.

DOCTOR STRANGE

MASTER OF THE MYSTIC ARTS

SPELLING MISTAKE

As a Master of the Mystic Arts, Doctor Strange knows all sorts of spells, though he struggles to contain his most recent one inside the Macchina di Kadavus.

Powerful Eye of Agamotto

Sling ring secured on belt

SUPER STATS

LIKES: Being right
DISLIKES: People who question his intelligence
FRIENDS: America Chavez, Wong
FOES: Mordo, Thanos
SKILLS: Opening portals, casting spells
GEAR: Sling ring, Cloak of Levitation

SET: 76205 Gargantos Showdown
YEAR: 2022

DID YOU KNOW?

Doctor Strange has formed a unique bond with his cloak. He is known to take great offense if anyone refers to it as a cape!

STRANGE BY NAME and strange by nature, Doctor Strange accepts that his actions across The Multiverse will not please everyone. Sporting the mysterious Eye of Agamotto, a portal-opening sling ring, and a sentient cloak, his minifigure is equipped for whatever strange adventures lie ahead.

OTHER STRANGES

STRANGE VARIANTS

THERE'S A DOCTOR STRANGE in every reality, but not all of them are the same. These minifigures have some similarities to "regular" Doctor Strange—but some very noticeable differences, too!

SINISTER STRANGE

Corrupted by the power of evil Darkhold magic, this Strange has a similar face to "regular" Strange but in a sickly yellow color.

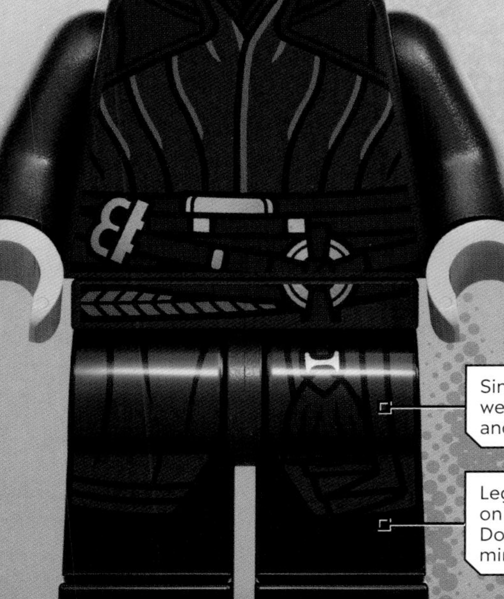

DEAD STRANGE

This undead Strange has an even more worrying skin tone, and his clothes are ripped and torn from when he was destroyed by a demon.

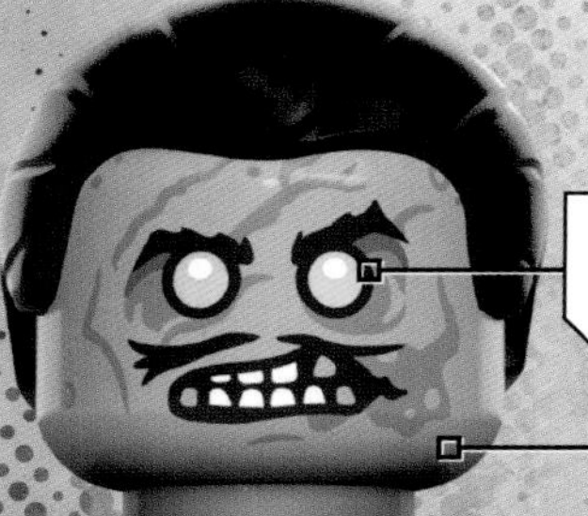

Tattered clothing

STAR-LORD

SPACE OUTLAW

PETER QUILL GREW UP as part of the Ravagers pirate clan, but he now embraces his role as leader of the Guardians of the Galaxy. The Guardians are no ordinary team, and Star-Lord is no ordinary leader. He pilots their spacecraft while cracking jokes and playing retro music.

Head piece printed with stubble and chin dimple

Guardians uniform based on the third *Guardians of the Galaxy* movie

Flight harness essential for space travel

Durable leather gloves

SUPER STATS

LIKES: Loud music
DISLIKES: Being bored
FRIENDS: Guardians of the Galaxy
FOES: Ronan, Thanos, The High Evolutionary
SKILLS: Piloting expertise, out-of-the-box thinking
GEAR: Helmet, Quad Blasters

SET: 76253 Guardians of the Galaxy Headquarters
YEAR: 2023

DID YOU KNOW?

This head piece is also used on the Owen Grady minifigures from LEGO® Jurassic World™—who are based on the same actor that plays Star-Lord, Chris Pratt.

ACE HELMET

Three earlier versions of Star-Lord's minifigure (including this one from 2014) came with a space helmet, which allowed Star-Lord to breathe in alien environments.

GAMORA

GREEN, MEAN MACHINE

WEAPON OF CHOICE
Gamora wields a collapsible sword. It was a gift from Thanos, though Gamora now uses it in battle against him!

Hair piece unique to Gamora

Leather coat as worn in Marvel Studios' *Avengers: Infinity War*

Studded leather combat vest

Coat details continue on legs

SUPER STATS

LIKES: Getting stuff done
DISLIKES: Looking silly
FRIENDS: Star-Lord, Nebula
FOES: Thanos
SKILLS: Combat, martial arts
GEAR: Sword

SET: 76107 Thanos: Ultimate Battle
YEAR: 2018

DID YOU KNOW?

Gamora and her adoptive sister, Nebula, have a complicated history. It's no wonder, then, that they appear in only one LEGO set together.

AN ASSASSIN with cybernetic upgrades, Gamora was on the side of her adoptive father, Thanos, before joining the Guardians of the Galaxy. Her battle skills come in handy as she tries to stop Thanos from finding the Infinity Stones.

ROCKET

FIERCE, FURRY GUARDIAN

ROCKET MIGHT BE THE FLUFFIEST member of the Guardians, but that doesn't stop him from blowing up everything in his path! Though his minifigure can be reckless, Rocket is always thinking—about a new plan, a new invention, or a new way to get rich! In the end though, he always looks out for his team.

Ravagers emblem on armored vest

Tail element attaches between torso and leg piece

SUPER STATS

LIKES: Sarcasm, exploding things
DISLIKES: Showing emotions
FRIENDS: Groot, Guardians of the Galaxy
FOES: The High Evolutionary, Ayesha, Thanos
SKILLS: Tech genius, piloting skills, incredible intellect
GEAR: None

SET: 76278 Rocket's Warbird vs. Ronan
YEAR: 2024

BABY ROCKET
A cute Baby Rocket figure comes in set 76254. Here, he escapes from his evil creator, The High Evolutionary.

GROOT

LITTLE GUARDIAN

Barklike skin is always growing

Moss grows on body

Zippered Ravagers costume in size XXS

TREELIKE GROOT JOINED the Guardians as an adult, but he chose to sacrifice himself to save the team. His minifigure is reborn as Baby Groot, with cute, short legs, an adorable childlike expression, and no memories from before!

DID YOU KNOW?

Groot has taken on many LEGO forms, including buildable figures and also minifigures and microfigures of Groot at different sizes.

SUPER STATS

LIKES: Saying "I am Groot!", dancing
DISLIKES: Anyone who attacks his friends
FRIENDS: Rocket, Guardians of the Galaxy
FOES: Ronan, Thanos, The High Evolutionary
SKILLS: Rapid body growth, super-strength, durability
GEAR: None

SET: 76282 Rocket & Baby Groot
YEAR: 2024

RARE SPECIES

Groot is a *Flora colossus*. His other minifigures are printed with the twisting vines, bark, and moss that make up his body.

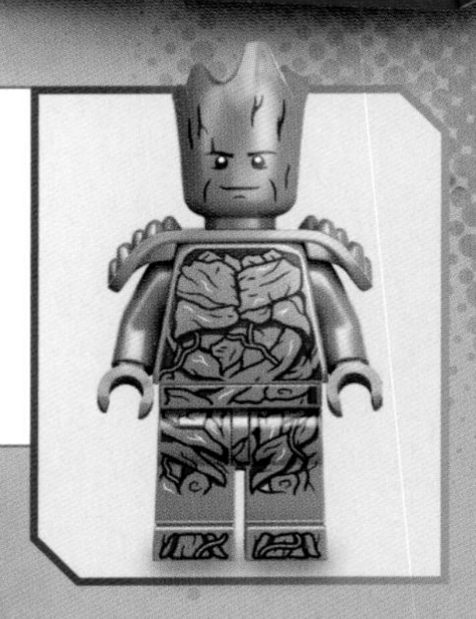

DRAX

THE DESTROYER

WHEN DRAX FIRST teamed up with the Guardians of the Galaxy, he was full of rage. Now he channels his anger for good, and his minifigure wears the red-and-blue Guardians uniform with pride. Loyal, fearless, and extremely strong, Drax joins his team on many heroic missions. Enemies beware!

Face markings feature on all Drax head pieces

Ravager flames feature on symbol

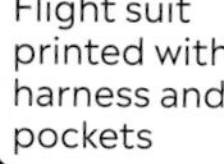

SUPER STATS

LIKES: Standing up for innocent people
DISLIKES: Dancing
FRIENDS: Mantis, Guardians of the Galaxy
FOES: Ronan the Accuser
SKILLS: Brute strength, healing factor
GEAR: Daggers

SET: 76255 The New Guardians' Ship
YEAR: 2023

MIGHTY BATTLE

Drax wields his famous dual daggers against Adam Warlock, who surprises the Guardians at their headquarters on Knowhere.

DID YOU KNOW?

Out of four Drax minifigures, two have gray skin and two green. The green skin is closer to Drax's comic-book appearance.

MANTIS

CALMING CREWMATE

INSECTOID MANTIS can sense and manipulate how others are feeling. This ability really bugs baddies when, in the midst of battle, Mantis makes them fall asleep, start dancing, or believe they are cute kittens. Even Thanos wishes she would buzz off!

Antennae are part of Mantis's unique hair piece

Bottle-green suit printing continues on back

Copper belt buckles are unique to this version of Mantis

Mantis uses touch to calm others

SUPER STATS

LIKES: Having friends
DISLIKES: Not being listened to
FRIENDS: Drax, Guardians of the Galaxy
FOES: Ego, Thanos
SKILLS: Sensing feelings, helping people fall asleep
GEAR: None

SET: 76193 The Guardians' Ship
YEAR: 2021

PART OF THE TEAM

Mantis had a lonely childhood, so she is thrilled to be one of the Guardians of the Galaxy. She joins her team in four LEGO sets, including on board their orange M-class ship.

CAPTAIN MARVEL

COSMIC HERO

CONSIDERED ONE OF THE most powerful beings in the universe, Captain Marvel got her powers from the Tesseract itself. Her mighty minifigure can fly through space, harness cosmic energy, and destroy Thanos's battleship all on her own. Talk about marvelous!

SUPER STATS

LIKES: Baseball, rock music
DISLIKES: Being deceived
FRIENDS: The Avengers, Nick Fury
FOES: Thanos, Kree
SKILLS: Cosmic energy, Air Force piloting skills, space flight, enhanced durability
GEAR: None

SET: 76237 Sanctuary II: Endgame Battle
YEAR: 2021

DID YOU KNOW?

Captain Marvel is the only Super Hero with powers gained from the Tesseract's Space Stone.

POWER BLASTS

Captain Marvel displays her skillful control of cosmic energy in set 76237. Her minifigure comes with lots of transparent orange "cosmic energy" accessories, which are sure to make Thanos run for cover!

SHE-HULK

ACE ATTORNEY

Hair piece in dark green color unique to She-Hulk

She-Hulk's Suit is durable and can adapt to change in body size

High-top white boots

DURING A CAR CRASH, Jen Walters is exposed to the blood of her cousin Bruce Banner. A high-flying attorney, Jen isn't too thrilled about the fact that she might turn into a green She-Hulk at any moment! Unlike Hulk, She-Hulk's LEGO form is not a big fig, but rather a bright green minifigure wearing Jen's knowing grin.

SUPER STATS

LIKES: Giving everyone a fair chance
DISLIKES: Makeup, gamma radiation
FRIENDS: Bruce Banner, Lulu
FOES: Titania
SKILLS: Super-strength, durability, confidence, intelligence
GEAR: She-Hulk's Suit

SET: 71039-13 LEGO Minifigures—Marvel Studios Series 2
YEAR: 2023

ALL IN THE DETAILS

She-Hulk's minifigure has plenty of amazing details, including arm printing and side printing on her legs. She carries a legal dossier and a cell phone, on which Wong is calling!

CHAPTER THREE

VILLAINS

WATCH OUT, there are villains about! Petty criminals feud with rival gangs, while robots, gods, and warlords plot to wipe out entire worlds. Earth's mightiest heroes stand up to these foes time and again. But after each victory, they must ask themselves: Who will attack next?

Thanos
Taskmaster
Loki

OBADIAH STANE

RUTHLESS CEO

DRESSED IN A SNAPPY SUIT, Obadiah Stane's minifigure looks every bit the suave businessman—though he is anything but! The double-crossing CEO of Stark Industries wants to take over the company—and get rid of Tony Stark while he's at it!

Bald head doesn't need a hair piece

Carefully groomed salt-and-pepper beard

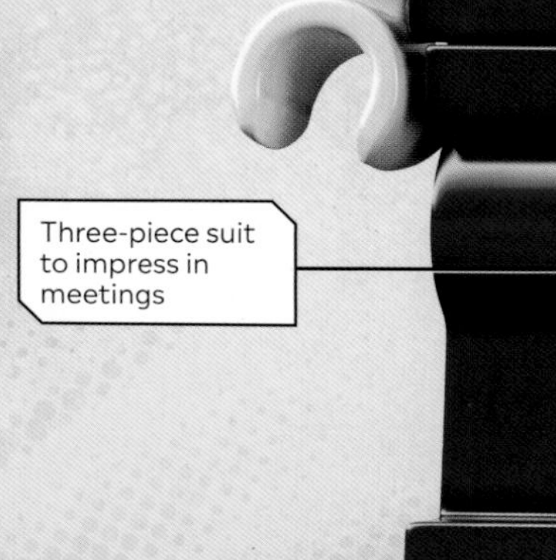

Flashy solid gold chain

Three-piece suit to impress in meetings

SUPER STATS

LIKES: Money
DISLIKES: Being second-in-command
FRIENDS: None
FOES: Tony Stark
SKILLS: Business sense, ambition, knowledge of Stark Industries
GEAR: Iron Monger suit

SET: 76190 Iron Man: Iron Monger Mayhem
YEAR: 2021

IRON MONGER
Stane's villainous suit is known as Iron Monger. The armored suit takes its inspiration from Iron Man, but it's much bigger—big enough to fit Stane's minifigure inside.

WHIPLASH

JEALOUS RIVAL

IVAN VANKO BELIEVES he should have everything Tony Stark has because their fathers used to be partners. So he creates an armored suit for his minifigure, powered by his own version of an Arc Reactor. Then he sets out to take on his nemesis, Iron Man.

DID YOU KNOW?

Ivan Vanko has based his armor on Iron Man's, but it can't compare. The LEGO® helmet doesn't even open in the same cool way!

Shiny, armored shoulder pads

Arc Reactor supplies energy to the cables

Silver armor dotted with scuff marks

Whiplash Mark 2 armor

Foot panels house anchor technology

SUPER STATS

LIKES: Revenge
DISLIKES: Tony Stark
FRIENDS: Justin Hammer
FOES: Iron Man, War Machine
SKILLS: Scientific knowledge, engineering skills
GEAR: Electrified whips

SET: 76216 Iron Man Armory
YEAR: 2022

ELECTRIFYING!
While Ivan Vanko's LEGO whips are perfectly safe, in the movie they are strong enough to stun even Iron Man!

ALDRICH KILLIAN

POWERFUL FOE

ALDRICH KILLIAN LOOKS like a regular CEO, and so does his minifigure—almost. The big clue that all is not right comes from his glow-in-the-dark head piece! A test subject for his own Extremis serum, Killian is a formidable foe due to healing abilities and fire-breathing powers.

Alternate face shows bulging red veins

Suave suit matches his CEO persona

Torso printed with button-down shirt details

SUPER STATS

LIKES: Science, power
DISLIKES: Failing
FRIENDS: Maya Hansen
FOES: Iron Man, War Machine
SKILLS: Extremis powers, genius mind
GEAR: None

SET: 76006 Iron Man: Extremis Sea Port Battle
YEAR: 2013

DID YOU KNOW?

While Killian's skin does not glow in the dark in the movie Marvel Studios' *Iron Man 3*, it does glow from within when he uses his fiery Extremis powers.

BOAT BATTLE

Killian's minifigure appears in just one LEGO set. There, he battles Iron Man and War Machine from a speedboat.

TREVOR SLATTERY

FAKE VILLAIN

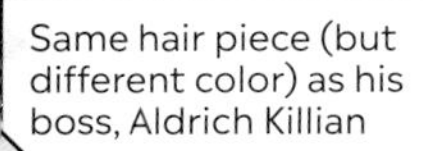

Same hair piece (but different color) as his boss, Aldrich Killian

Head piece unique to Slattery

Green-and-gold robes continue from torso to legs

TREVOR SLATTERY IS a below-average actor, but his minifigure is dressed for the role of a lifetime! His invented costume is of an imaginary villain whom Slattery pretends to be under orders from Aldrich Killian. Slattery turns out to be anything but threatening when the truth is revealed!

SUPER STATS

LIKES: Fame
DISLIKES: Facial injuries
FRIENDS: Aldrich Killian
FOES: Tony Stark
SKILLS: Acting, translation skills
GEAR: Audio visual equipment

SET: 76007 Iron Man: Malibu Mansion Attack
YEAR: 2013

FAKE NEWS
An Extremis soldier flies Slattery to Tony Stark's Malibu Mansion in set 76007. Here, the actor records yet another fake video for the press.

HYDRA AGENT

HATEFUL HENCHMAN

IN A SOKOVIAN FOREST, hordes of Hydra agents defend their base from the Avengers. Despite his confident smirk, this minifigure is about to discover just how powerful the punch of an angry Hulk can be!

Hydra emblem also printed on back

Utility pouch carries spare ammunition and sometimes snacks

Heavy body armor hinders movement

Snow camouflage ideal for the icy Sokovian landscape

SUPER STATS

LIKES: Warm hats
DISLIKES: Being cold
FRIENDS: Hydra, Baron Strucker
FOES: Avengers
SKILLS: Close combat, obedience
GEAR: Chitauri weapons

SET: 76189 Captain America and Hydra Face-Off
YEAR: 2021

FULL UPGRADE

Some Hydra soldiers are upgraded with alien Chitauri technology, which includes cybernetic implants and armor.

DID YOU KNOW?

There are two other Hydra minifigures, who wear green based on their Marvel Comics look. One comes dressed in a diving suit.

ULTRON

METAL MONSTER

Remnant of Iron Legion helmet

Worn Avengers emblem

Scrap metal forms body

Spare parts from lab

DID YOU KNOW?

Ultron tried to build himself a body of vibranium fused with living tissue, but the Avengers got to it first. It became the Super Hero Vision.

A BEING OF artificial intelligence, Ultron wants complete global destruction—and also to defeat Iron Man. His minifigure has several versions, reflecting Ultron's quest to create the perfect body. This 2023 minifigure displays Ultron's first body, which was cobbled together from scraps of metal in the Avengers' lab.

SUPER STATS

LIKES: Growing stronger
DISLIKES: Weak, foolish humans
FRIENDS: Ultron Sentries
FOES: Iron Man, Vision, Avengers, humankind
SKILLS: Vast knowledge, flight, repulsor technology
GEAR: Chitauri weapons

SET: 76269 Avengers Tower
YEAR: 2023

OUT OF CONTROL

Ultron was created by Tony Stark using data from the Mind Stone. Stark thought Ultron would be part of a peacekeeping program, but Ultron's army ends up wreaking havoc!

CROSSBONES

ANGRY ENEMY

FORMER S.H.I.E.L.D. AGENT (but really a Hydra spy) Brock Rumlow fooled everyone for a while. But his new name Crossbones makes his evil intentions crystal clear. His minifigure wears a scary metal mask and an armored bodysuit, instilling fear into everyone he meets. His only objective is to make the Avengers pay for defeating him.

Mask hides old battle scars

Gauntlets contain hidden weapons

Armored suit has a built-in detonator

SUPER STATS

LIKES: Revenge, revenge, revenge
DISLIKES: Falling buildings
FRIENDS: None
FOES: Black Widow, Captain America, Falcon
SKILLS: Street fighting, knowledge of S.H.I.E.L.D.
GEAR: Grenades, armored suit

SET: 76050 Crossbones' Hazard Heist
YEAR: 2016

TRUCK TAKEDOWN
Crossbones lures the Avengers into battle. He fights them from an armored truck, but it's no match for Earth's Mightiest Heroes!

THE COLLECTOR

HOLDER OF TREASURES

TANELEER TIVAN IS PROUD to be known as the Collector. His museum houses the largest collection of items in the universe (though it's best not to ask how he obtained most of them). It seems fair that his minifigure is a collectible item in itself: only a few copies were produced, as a Comic-Con giveaway in 2014.

SUPER STATS

LIKES: Rare and valuable objects
DISLIKES: Common things
FRIENDS: None
FOES: Thanos
SKILLS: Dealing and bargaining, enhanced durability, near-immortality
GEAR: Whatever treasures his collection holds

SET: Comcon035 The Collector
YEAR: 2014

ADDING TO THE COLLECTION

Living up to his name, the Collector wants to collect the Infinity Stones. He tries to buy the Power Stone from the Guardians of the Galaxy, but the deal falls through.

RONAN
THE ACCUSER

A BAD-TEMPERED Kree warlord, Ronan's minifigure battles the Guardians of the Galaxy in two LEGO sets. He aims to steal the Power Stone for Thanos but, after trying to take it for himself, ends up getting destroyed by it. Oops.

War paint applied by his servants

Kree armor enhances Ronan's natural super-strength

War apron print continues on legs

SUPER STATS

LIKES: Holding a grudge
DISLIKES: Powerful mortals
FRIENDS: Thanos
FOES: Skrulls, Guardians of the Galaxy, Captain Marvel
SKILLS: Super-strength, durability
GEAR: Cosmi-Rod, the Orb

SET: 76278 Rocket's Warbird vs. Ronan
YEAR: 2024

DID YOU KNOW?

With only one minifigure available in one LEGO set since 2014, Ronan has been a rare, collectible item. The release of a new minifigure in 2024 got fans all excited!

WARLORD AT WAR
The Ronan minifigure from 2014 (set 76021) wears Kree armor, too. But this one has a slightly different design.

TASERFACE

RAVAGERS TRAITOR

Hair piece with topknot is also worn by Valkyrie

A MEMBER OF THE Ravagers pirate clan, Taserface grows dissatisfied with his leader, Yondu. So he does what all villains do—takes power for himself! His minifigure wears a Ravagers suit that is partially covered by his unkempt beard, his purple face glaring at any Ravager who would dare disobey him.

DID YOU KNOW?

When Rocket questions and makes fun of Taserface's name, the Ravager answers that "It's metaphorical!"

Unique leg and torso pieces make Taserface a popular minifigure

SUPER STATS

LIKES: Being a Ravager
DISLIKES: People mocking his name
FRIENDS: The Ravagers
FOES: Yondu, Rocket Raccoon, Groot
SKILLS: Combat, piracy
GEAR: Ravager weapons

SET: 76079 Ravager Attack
YEAR: 2017

IT'S A MUTINY!
Taserface comes in only one LEGO set, which depicts the moment he leads the Ravager rebellion and takes control of the clan.

AYESHA

SOVEREIGN HIGH PRIESTESS

REMOTE CONTROL
The Sovereign's battle fleet is made up of drones. Ayesha and her warriors control their mechanical army from afar.

DID YOU KNOW?

It's rare for every part of a minifigure to be unique. However, Ayesha is made of all unique elements, from her golden hair to the boots of her Sovereign suit.

Pearl-gold hair piece is unique to Ayesha

Gold details emphasize the Sovereign's wealth

Space-resistant protective fabric

SUPER STATS

LIKES: Feeling superior to others
DISLIKES: Being disrespected
FRIENDS: The High Evolutionary, Adam Warlock
FOES: Guardians of the Galaxy
SKILLS: Genius tactical mind
GEAR: Drone army

SET: 76080 Ayesha's Revenge
YEAR: 2017

AS HIGH PRIESTESS of the Sovereign (an advanced people created by The High Evolutionary), Ayesha is proud but also rather rude to others. Her minifigure wears a Sovereign battlesuit, though Ayesha prefers to wear her all-gold High Priestess robes.

ADAM WARLOCK

SOVEREIGN SON

SECOND-IN-COMMAND to Ayesha, Adam Warlock is a proud Sovereign leader and warrior. While his behavior can be childish, Adam's powerful minifigure is strong, durable, and confident enough to take on the Guardians of the Galaxy by himself.

Skull used as cape clasp

Red cape also worn by The Scarlet Witch

Gold Sovereign battle armor

SUPER STATS

LIKES: His mom
DISLIKES: Being bossed around
FRIENDS: Ayesha
FOES: Guardians of the Galaxy (though he later joins them)
SKILLS: Energy blasts, space flight, healing ability
GEAR: None

SET: 76255 The New Guardians' Ship
YEAR: 2023

SPACE MAN

Adam's Sovereign DNA has been genetically perfected. This means his body is incredibly strong and can survive in the cold vacuum of space.

LOKI

GOD OF MISCHIEF

THE ADOPTED SON of Asgard's King Odin, Loki has always had lofty aspirations. He wants to rule—Asgard, Earth, wherever—which often sees him at odds with his more noble brother, Thor. Loki's minifigure usually wears stylish black-and-green Asgardian clothing and an irritating smirk.

Large, golden horned headdress

Fine Asgardian leather cloak makes everything more dramatic

SUPER STATS

LIKES: Having an audience
DISLIKES: Lightning
FRIENDS: It's complicated
FOES: It's also complicated
SKILLS: Creating illusions
GEAR: Scepter

SET: 76248 The Avengers Quinjet
YEAR: 2023

DID YOU KNOW?

Loki is known for talking. A lot. This minifigure's alternate face shows him with a mask over his mouth to prevent him from speaking and lying any more.

CHANGING TIMES

Loki's Time Variance Authority uniform is much more practical than his usual choice of attire. Based on the Marvel Studios' *Loki* series, this minifigure wears a TVA uniform with the word "Variant" printed on the back.

HELA

GODDESS OF DEATH

Horned headdress unique to Hela

Green war paint on helmet

THOR NEVER KNEW he had an older sister, and when he finds out, he wishes it wasn't true. Hela's minifigure radiates evil, from her grim expression and sinister headdress, and in her desire for destruction. Unfortunately, as Thor soon discovers, Hela is very hard to destroy!

Hela can generate her costume at will

Dark green cape matches suit

SUPER STATS

LIKES: Death
DISLIKES: Being banished
FRIENDS: Fenris, Skurge (but not really)
FOES: Thor, Valkyrie
SKILLS: Summoning weapons at will, increased durability
GEAR: Cool helmet

SET: 76084 The Ultimate Battle for Asgard
YEAR: 2017

ARMY OF THE DEAD
True to her title, Hela resurrects Asgardian warriors from the dead to storm Asgard. She also summons her long-gone hound, Fenris.

GRANDMASTER

CROWD-PLEASER

THE GRANDMASTER RULES the planet of Sakaar, where he entertains the masses—and himself—by pitting warriors against each other in a tough tournament known as the Contest of Champions. Always smiling, his minifigure cares nothing for the plight of the captive warriors.

Blue makeup adorns eyes and chin

Both sides of his face feature a smug grin

Vibrant red silk lining

Lavish patterned golden robes

Pants made from luxurious, stretchy fabric

SUPER STATS

LIKES: Creating an entertaining spectacle
DISLIKES: Revolutions
FRIENDS: Topaz
FOES: Thor, Hulk
SKILLS: Commands authority, smooth manner
GEAR: Extravagant clothing

SET: 76088 Thor vs. Hulk: Arena Clash
YEAR: 2017

ALL EYES ON ME!
Sitting on his LEGO throne, there's nothing the Grandmaster loves more than being in the spotlight, far away from any hint of danger.

GORR

SUMMONER OF SHADOWS

DID YOU KNOW?

Gorr's minifigure is based on his appearance in the Marvel Studios' movie *Thor: Love and Thunder*.

DRAPED IN SIMPLE white robes, Gorr is on a violent mission: to destroy every single god. This brings him face-to-face with Thor, the God of Thunder, and Thor isn't about to let anyone defeat him!

SUPER STATS

LIKES: Revenge
DISLIKES: Gods
FRIENDS: Shadow monsters
FOES: Thor, The Mighty Thor
SKILLS: Immortality
GEAR: Necrosword

SET: 76207 Attack on New Asgard
YEAR: 2022

SHADOWY SURPRISE

Gorr harnesses a powerful weapon, the Necrosword. He uses it to summon scary shadow monsters!

THANOS

TRAVELING WARLORD

ALWAYS PURPLE AND always angry, Thanos is on a mission to wipe out half the population of the universe. At first he takes it one planet at a time, but he soon realizes a quicker way is within his grasp. Collecting the six Infinity Stones will let him do it in a snap.

SUPER STATS

LIKES: Infinity Stones
DISLIKES: Overpopulation
FRIENDS: None
FOES: The Avengers, Guardians
SKILLS: Immense strength, durability
GEAR: Infinity Gauntlet

SET: 76266 Endgame Final Battle
YEAR: 2023

GAUNTLET GRAB
With a snap of his purple fingers, Thanos could use the Infinity Gauntlet to remove half of all life. No wonder Iron Man and the Avengers are so keen to get hold of it!

CHITAURI & OUTRIDERS

THANOS'S SOLDIERS

THANOS ATTACKS EARTH using armies from other planets. The Chitauri and Outriders are alien species that don't think for themselves, making them perfect as obedient foot soldiers. Though Thanos commands thousands of these aliens, there are only four Chitauri and five Outrider minifigure variants.

Head piece printed with huge gold teeth, but no eyes

Extended arm with long claws on the end

Gold armor protects thick, gray skin

CHITAURI

These reptilian aliens attack New York City through a portal in the sky, causing vast destruction. But one good thing comes from this battle—the Avengers assemble for the first time ever!

OUTRIDERS

The Outriders are deployed for battle on the plains of Wakanda. They have multiple limbs and scarily sharp teeth. Some Outrider minifigures come with four extra arms!

Electronic networks are built into body

DID YOU KNOW?

Chitauri are humanoid in shape, whereas Outriders are more like beasts, running on all fours. However, in LEGO sets, they both come in minifigure form.

EBONY MAW

THANOS'S REPRESENTATIVE

Wrinkled face with scowling expression

ONE OF THANOS'S MOST trusted warriors, Ebony Maw is on the hunt for the Infinity Stones. He shows an impressive use of telekinesis in battle, though his minifigure's face is set in an eternal scowl!

DID YOU KNOW?

Ebony Maw is part of the Children of Thanos, a group of soldiers raised by Thanos. They consider themselves his adopted children.

Central panel adds grandeur

Protective armor printed on front and back of torso

SUPER STATS

LIKES: Being Thanos's favorite
DISLIKES: Anyone who says a bad word about Thanos
FRIENDS: Thanos, Cull Obsidian
FOES: Doctor Strange, Iron Man
SKILLS: Telekinesis
GEAR: None needed

SET: 76218 Sanctum Sanctorum
YEAR: 2022

SANCTUM ATTACK
There are two Ebony Maw minifigure versions, which come in different LEGO sets. Both sets feature Maw attacking the Sanctum Sanctorum.

CULL OBSIDIAN

BRUTE OF THANOS

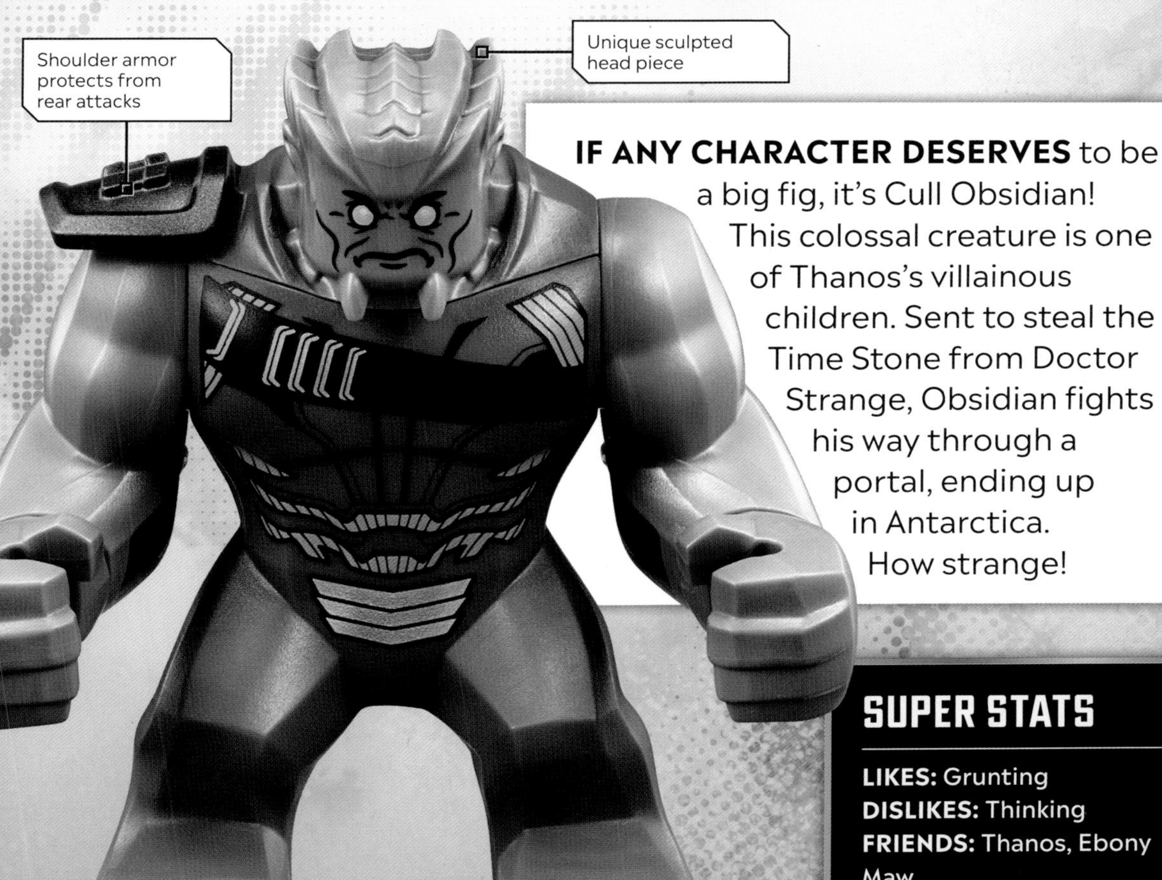

IF ANY CHARACTER DESERVES to be a big fig, it's Cull Obsidian! This colossal creature is one of Thanos's villainous children. Sent to steal the Time Stone from Doctor Strange, Obsidian fights his way through a portal, ending up in Antarctica. How strange!

LIKES: Grunting
DISLIKES: Thinking
FRIENDS: Thanos, Ebony Maw
FOES: Hulk, Doctor Strange, Iron Man
SKILLS: Strength, durability, healing
GEAR: Chain hammer

SET: 76108 Sanctum Sanctorum Showdown
YEAR: 2018

DEFEATED WARRIORS
Obsidian finds himself in a tangle along with Ebony Maw during a battle against the Avengers at the Sanctum Sanctorum.

PROXIMA MIDNIGHT

WARRIOR OF THANOS

PROUD TO SERVE HER MASTER, THANOS, Proxima Midnight sets off in search of the Mind Stone. Her minifigure has both an angry face and a smirking expression, which she wears to taunt her enemies on the battlefield.

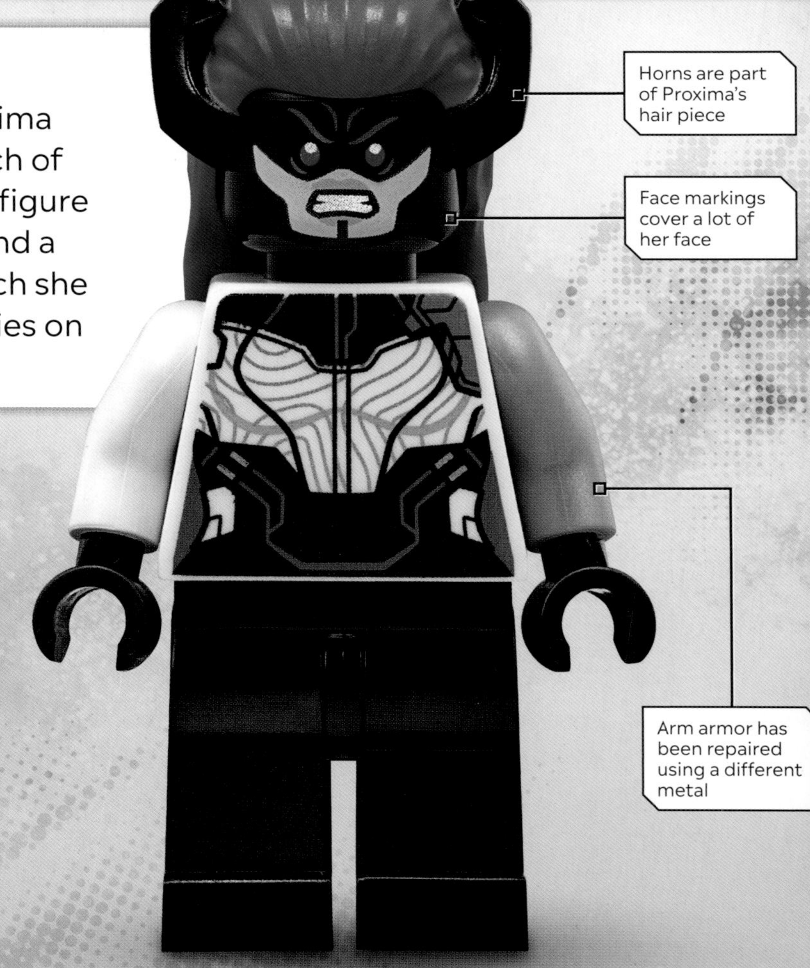

Horns are part of Proxima's hair piece

Face markings cover a lot of her face

Arm armor has been repaired using a different metal

SUPER STATS

LIKES: Defeating her enemies
DISLIKES: Disappointing Thanos
FRIENDS: Thanos, Corvus Glaive
FOES: Vision, Wanda Maximoff, Hulk
SKILLS: Super-strength, agility, speed
GEAR: Spear that fires energy blasts

SET: 76104 The Hulkbuster Smash-Up
YEAR: 2018

RACE FOR THE STONE
Proxima battles the Avengers for the precious and powerful Mind Stone. She must succeed—or face Thanos's wrath.

CORVUS GLAIVE

SOLDIER OF THANOS

Yellow teeth give him a ghastly look

Intricate armor with metallic gold details

Tattered cloak has been through many battles

ANOTHER OF THANOS'S RUTHLESS CHILDREN, Corvus Glaive accompanies Proxima Midnight to Europe and then Wakanda to take the Mind Stone from Vision. Glaive is not a natural smiler, so his minifigure's scary-looking grin must mean that the mission is going well!

SUPER STATS

LIKES: Completing missions
DISLIKES: Being defeated by enemies he thinks are weak
FRIENDS: Thanos, Proxima Midnight
FOES: Vision, Captain America
SKILLS: Super-strength, durability, speed
GEAR: Glaive sword

SET: 76103 Corvus Glaive Thresher Attack
YEAR: 2018

WAKANDAN DEFEAT

Glaive's arrogance proves to be his downfall when he turns his back on Vision during a battle in Wakanda. Big mistake!

ERIK KILLMONGER

PANTHER RIVAL

N'JADAKA BELIEVES THAT he should be the ruler of Wakanda, so he challenges his cousin T'Challa for the Black Panther mantle. Known as Killmonger for his violent past, N'Jadaka's battle-ready minifigure puts up a good fight but is defeated by his more powerful cousin.

Ancient mask stolen from a museum

DID YOU KNOW?

This highly detailed ritual mask was created especially for Erik Killmonger's minifigure.

Battle vest worn over combat clothes

Cargo pockets hold ammunition

SUPER STATS

LIKES: Dangerous missions
DISLIKES: Oppression
FRIENDS: Ulysses Klaue
FOES: Black Panther, Nakia
SKILLS: Navy SEAL training
GEAR: Black Panther suit

SET: 76100 Royal Talon Fighter Attack
YEAR: 2018

PANTHER BATTLE
Erik Killmonger's other minifigure also wears the Black Panther suit, but with unique golden accents. He draws upon its power to fight against T'Challa.

ULYSSES KLAUE

VIBRANIUM THIEF

ULYSSES KLAUE IS A MASTER CRIMINAL. His minifigure prides himself on being the only outsider that ever saw the precious vibranium of Wakanda—and got out alive! He sells stolen vibranium to Ultron and teams up with Erik Killmonger. But joining forces with villains never ends well. Klaue would agree, if he still could!

Unique head piece has two expressions: angry and angrier

Neck branded with Wakandan symbol for "thief"

Prosthetic arm conceals weapon inside

Shirt and vest to give the appearance of a respectable businessman

SUPER STATS

LIKES: Money, jazz music
DISLIKES: Cuttlefish
FRIENDS: Ultron, Erik Killmonger
FOES: Black Panther, the Avengers
SKILLS: Smuggler, thief, knowledge of the criminal world
GEAR: Sonic arm cannon, store of vibranium

SET: 76100 Royal Talon Fighter Attack
YEAR: 2018

SECRET WEAPON

Klaue lost an arm during his dealings with Ultron. So he built himself a prosthetic one that can transform into a sonic cannon!

KING NAMOR

RULER OF TALOKAN

KING NAMOR IS FIERCELY protective of his underwater kingdom, Talokan, and its vibranium resources. His bold minifigure wants to unite with Wakanda against the rest of the world, but when Shuri refuses, Namor wages war.

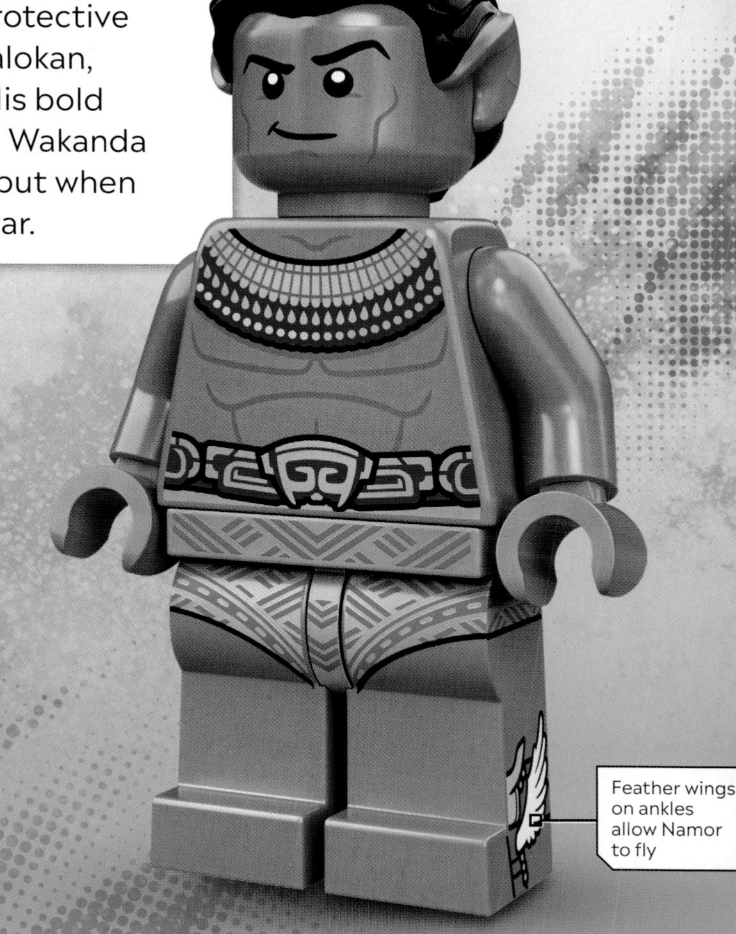

Feather wings on ankles allow Namor to fly

SUPER STATS

LIKES: Defending his people and culture
DISLIKES: Surface people
FRIENDS: Attuma
FOES: Shuri, Ironheart
SKILLS: Can breathe underwater and above water, flight, slow aging
GEAR: Vibranium spear

SET: 76213 King Namor's Throne Room
YEAR: 2022

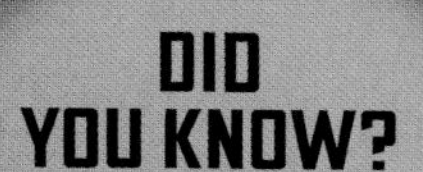

DID YOU KNOW?

Every piece of Namor's minifigure is unique—from his hair piece with connected ears to his printed legs.

JAW-SOME THRONE

Namor rules from an underwater throne made out of a shark's jawbone and decorated with seaweed. He is proud of Talokan's rich history.

ATTUMA

WARRIOR OF TALOKAN

Massive helmet with dual molded spikes

A LEADER OF THE Talokan army, Attuma captures Black Panther under King Namor's orders. The Talokanil are an underwater people, and Attuma's minifigure reflects this heritage: his headdress and clothing are made from shark hide and fish bones!

DID YOU KNOW?

Like that of his ruler, King Namor, Attuma's minifigure is made from unique parts.

Waistband with buckle

SUPER STATS

LIKES: Battles
DISLIKES: Threats to his kingdom
FRIENDS: Namor
FOES: Okoye, Shuri, Ironheart
SKILLS: Combat
GEAR: Spear, respirator

SET: 76211 Shuri's Sunbird
YEAR: 2022

UNDERWATER ATTACK

Attuma is a skilled warrior, both in and out of water. He wields a Talokanil vibranium spear.

YELLOWJACKET

ANT-MAN'S ENEMY

DARREN CROSS WAS A STUDENT of Hank Pym, who created the Ant-Man suit. But Cross stole the technology for himself and created a more advanced suit, which his minifigure wears as Yellowjacket. Not only can Yellowjacket grow or shrink at will, but he can fly and fire energy beams from his back-mounted limbs.

Unique helmet mold with transparent yellow visor

Harness attaches to back-mounted limbs

Yellowjacket Suit made from Kevlar

SUPER STATS

LIKES: Being powerful
DISLIKES: Anyone who dares question him
FRIENDS: Hydra
FOES: Ant-Man, Hank Pym
SKILLS: Size manipulation, enhanced strength, flight
GEAR: Yellowjacket Suit with energy beams

SET: 76039 Ant-Man Final Battle
YEAR: 2015

BUG BATTLE
Shrunk down to the size of a wasp, Yellowjacket fights Ant-Man, who rides into battle on his ant friend, Ant-thony.

GHOST

PHANTOM FOE

Extra eyes provide better vision while phasing

Phase-shifting panels printed on Ghost's armor

AVA STARR IS KNOWN as Ghost, and not just because of her scary mask. As a result of a quantum accident she can turn invisible and pass through solid objects in a very ghostly way. Ghost needs quantum energy to survive, which sets her on a path against Ant-Man and The Wasp.

SUPER STATS

LIKES: Teddy bears
DISLIKES: Pain
FRIENDS: Bill Foster
FOES: Ant-Man, The Wasp, S.H.I.E.L.D.
SKILLS: Phasing through solid objects, invisibility, espionage and stealth abilities
GEAR: Ghost Suit

SET: 76109 Quantum Realm Explorers
YEAR: 2018

A GHOSTLY FACE

Ghost's suit and mask help her to control her powers. But when she chooses to reveal her identity, her minifigure's head piece also shows Starr's true face.

BARON KARL MORDO

ALTERNATE SORCERER SUPREME

WHEN DOCTOR STRANGE travels across The Multiverse, he has a surprise meeting with Karl Mordo, a former friend. This Mordo is his reality's Sorcerer Supreme—and he hates Doctor Strange! Mordo's minifigure wears pale green Sorcerer Supreme robes and an impatient sneer.

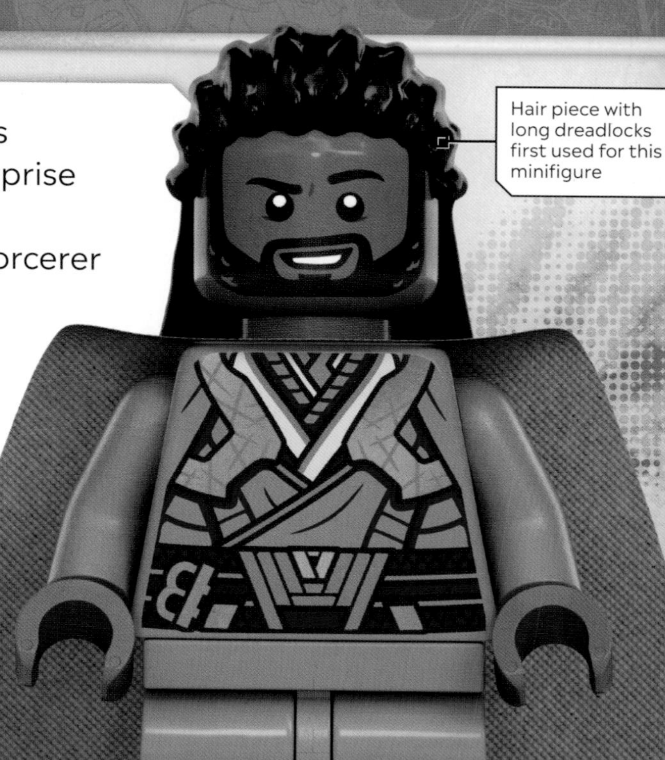

Hair piece with long dreadlocks first used for this minifigure

Sorcerer Supreme robes include a long, green cloak in this reality

SUPER STATS

LIKES: Following the rules
DISLIKES: Changing his mind
FRIENDS: The Illuminati
FOES: Doctor Strange
SKILLS: Mystic arts, martial arts
GEAR: Sling ring, magically enhanced boots, magic sword

SET: 76218 Sanctum Sanctorum
YEAR: 2022

ANOTHER MORDO
The Karl Mordo that Doctor Strange knew helped recruit and train him as a skillful Master of the Mystic Arts.

TASKMASTER

MYSTERIOUS ASSASSIN

DID YOU KNOW?

There are two versions of Taskmaster's minifigure. The only difference between them is that one comes with a hood.

High-tech goggles with digital display

Skull-like face partially hidden beneath hood

Torso printed with wiring and circuitry

TASKMASTER is a warrior sent to hunt down Black Widow as revenge for past events. This sinister assassin can mimic the moves of any opponent, which gives a real advantage in battle. Black Widow, beware!

SUPER STATS

LIKES: Revenge
DISLIKES: Humor
FRIENDS: General Dreykov
FOES: Black Widow, Yelena Belova
SKILLS: Mimicking fighting styles
GEAR: Shield, sword, mechanical backpack

SET: 76162 Black Widow's Helicopter Chase
YEAR: 2020

HELICOPTER CHASE
Taskmaster pilots a powerful armed helicopter in pursuit of Black Widow in set 76162.

WENWU

LEADER OF THE TEN RINGS

WENWU IS AN international warlord and leader of The Ten Rings criminal group. He possesses a set of 10 mystical iron rings, which give him immense powers, including super-strength and immortality. Wenwu has risen to power through the ages and is thousands of years old—though his minifigure doesn't look it!

Stern expression makes everyone cower

Ten Rings battlesuit fuses traditional robes with reinforced armor

Ornate printing continues on legs

SUPER STATS

LIKES: Power
DISLIKES: Moving forest mazes
FRIENDS: Razor Fist, The Ten Rings
FOES: Shang-Chi, Xialing
SKILLS: Immortality, super-strength, power blasts, combat skills
GEAR: The Ten Rings

SET: 76176 Escape from The Ten Rings
YEAR: 2021

IN THE RING
The Ten Rings are not just a source of power and immortality to Wenwu—he also wields them in battle as flexible, transforming, energy-blasting weapons!

RAZOR FIST

TEN RINGS ASSASSIN

DID YOU KNOW?

Razor Fist's head piece has two expressions: one scowling and one smiling. This is because when he joins forces with Shang-Chi and Xialing he becomes much more friendly!

THIS TRAINED ASSASSIN is so proud of his unique weapon—a blade in place of his hand—that he uses it for his nickname, Razor Fist. His minifigure is Wenwu's most loyal warrior, sent to find his boss's children, Shang-Chi and Xialing.

Pocketed vest has Ten Rings logo printed on back

Combat vest allows freedom of movement

Right hand transforms into a sharp weapon (by attaching a LEGO blade element)

SUPER STATS

LIKES: Being the strongest person in the room
DISLIKES: People stealing his car
FRIENDS: Wenwu, The Ten Rings
FOES: Shang-Chi, Xialing, Katy
SKILLS: Combat and martial arts prowess, strength
GEAR: Cybernetic arm blade

SET: 76176 Escape from The Ten Rings
YEAR: 2021

FULLY ARMED
In set 76176, Shang-Chi battles hard-to-escape-from members of The Ten Rings. Razor Fist uses his arm-blade to try to stop him!

DEATH DEALER

TEN RINGS WARRIOR

DOUBLE TROUBLE
Death Dealer uses his two signature daggers to threaten Xialing in battle, but Wenwu's daughter shows no fear!

Striking, scary mask provided by Wenwu

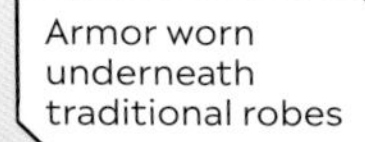

Exploding darts held on belt

DID YOU KNOW?

Echoing their roles in the Marvel Studios' movie *Shang-Chi and The Legend of The Ten Rings*, the actor who plays Death Dealer (Andy Le) helped train Simu Liu, who plays Shang-Chi.

SUPER STATS

LIKES: Being the best
DISLIKES: Lazy students
FRIENDS: Wenwu, Razor Fist
FOES: Shang-Chi
SKILLS: Martial arts expertise
GEAR: Twin daggers

SET: 76177 Battle at the Ancient Village
YEAR: 2021

HIGHLY SKILLED IN martial arts, Death Dealer trains the warriors of The Ten Rings criminal gang. His minifigure wears a unique costume and mask that makes him stand out from other The Ten Rings members. It certainly makes him look intimidating in battle!

TALOS

SKRULL ALLY

Unique head print shows grumpy Skrull face

Dark purple leather coat

A SKRULL SHAPE-SHIFTER, Talos can take on the appearance of anyone he chooses. He even impersonates Nick Fury for a while! But his minifigure has only ever appeared in one form: his original body. Talos works to save the Skrulls in secret, and he shows himself in only one LEGO set.

SUPER STATS

LIKES: Loyalty
DISLIKES: Flerken
FRIENDS: Nick Fury, Captain Marvel
FOES: Kree
SKILLS: Shape-shifting
GEAR: None

SET: 76127 Captain Marvel and The Skrull Attack
YEAR: 2019

CUTE CAT?
While most people think Goose is a cute cat, Talos isn't fooled. And he is proven right—Goose is actually a dangerous Flerken creature!

COMIC-BOOK VILLAINS

ANOTHER ROGUES' GALLERY

THERE'S NO SHORTAGE of villains that seek to rule or destroy the world—or cause other mischief of all kinds. From henchmen to mutants to cyborgs, these evil minifigures always find a way to bring the battle to LEGO's mightiest heroes!

Orange visor appears on all AIM Agent head pieces

Armored uniform with AIM insignia

AIM AGENT
Dressed in their yellow-and-black uniforms, AIM agents fight the Avengers often. AIM minifigures appear in seven LEGO sets, and they are all faceless henchmen armed with various weapons.

RED HULK
One of Hulk's strongest comic-book foes, Red Hulk has similar powers to the green Avenger and his red body also radiates intense heat! His angry big fig causes destruction in just one LEGO set.

Hair is attached to the single piece that makes up the big fig's head, torso, and legs

Same hair piece as Gamora but with different color streaks

RED SHE-HULK
The daughter of Red Hulk, Red She-Hulk joins her dad to take on Hulk and his cousin, She-Hulk. It's no surprise that the ensuing battle is fierce and furious!

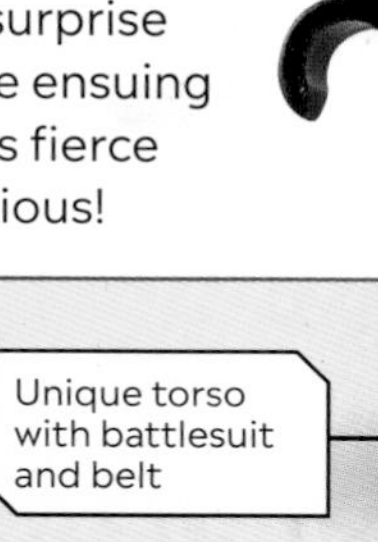

Unique torso with battlesuit and belt

SUPER-ADAPTOID

This villain has the useful power to take on the abilities of his enemies—which explains why his minifigure looks like a combination of Iron Man, Falcon, and Thor!

DID YOU KNOW?

M.O.D.O.K.'s figure is not a minifigure or a big fig. Like the character, his LEGO form is unique, and it is built from 123 elements!

HYPERION

Banished from his home planet, Hyperion travels to Earth and causes trouble there! His minifigure joins Thanos in an attack against the Avengers in space.

M.O.D.O.K.

M.O.D.O.K. used to be an ordinary man until an experiment went wrong. Now he's a genius who has a huge head to house his oversize brain and a small, useless body.

CHAPTER FOUR

ALLIES

WITH SUCH GREAT POWERS, Super Heroes shoulder a lot of responsibility. But they don't have to go it alone. Many brave allies join the fight against evil, using their unique skills, gadgets, or even quirky personalities. Some even go on to become friends and teammates.

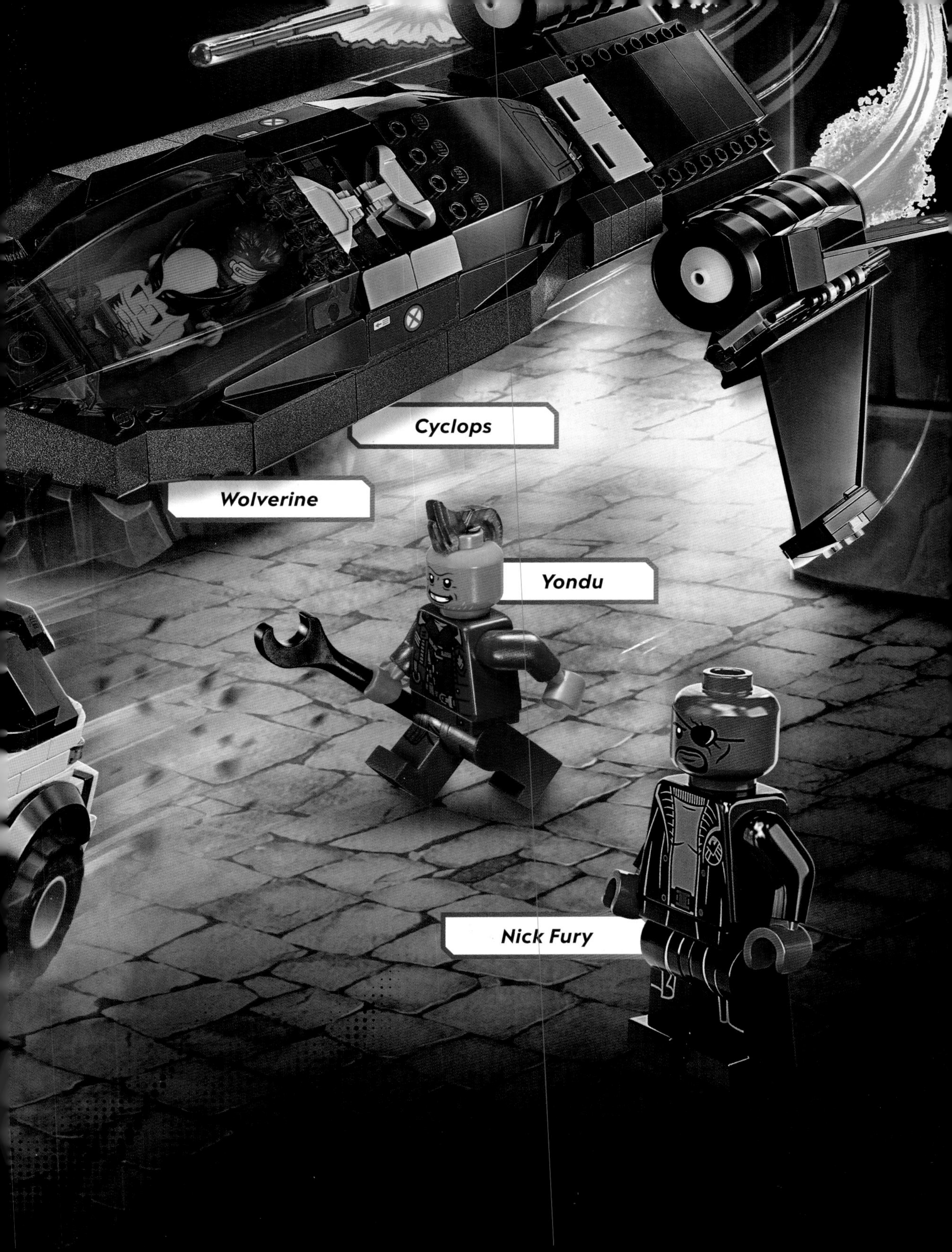
Cyclops
Wolverine
Yondu
Nick Fury

HAPPY HOGAN

STARK'S SECURITY GUY

HEAD OF SECURITY at Stark Industries, Happy Hogan is more than a bodyguard. He is brave, strong, and always alert—but also one of Tony Stark's best friends. His minifigure appears in only one set and, in keeping with his role as a security agent, blends into the background in a nondescript black suit.

DID YOU KNOW?

Happy is not this minifigure's real name. Stark gave him the ironic nickname "Happy" because he's not very smiley.

Crisp suit is Hogan's standard attire

Muscular arms from professional boxing

SUPER STATS

LIKES: Boxing, watching TV
DISLIKES: The paparazzi
FRIENDS: Tony Stark, May Parker, Spider-Man
FOES: Ivan Vanko, Vulture, Mysterio
SKILLS: Boxing, security protocols
GEAR: None

SET: 76130 Stark Jet and the Drone Attack
YEAR: 2019

PILOT PAL
Hogan's many skills include piloting the Stark Jet. He keeps his cool during battle, even when Spider-Man tries to discuss Happy's relationship with May!

PEPPER POTTS

STARK INDUSTRIES CEO

Freckles printed on head piece

Button-down shirt beneath tailored jacket

No-nonsense business suit for CEO job

PEPPER POTTS WENT from being Tony Stark's assistant to being the CEO of Stark Industries, eventually becoming a hero. Pepper's minifigure is proud of her achievements: running a multinational business, defeating villains, and—most importantly—dealing with Stark's constant sarcasm and silly jokes!

SUPER STATS

LIKES: Art, birthday presents
DISLIKES: Job hunting, strawberries
FRIENDS: Tony Stark, Happy Hogan, Phil Coulson
FOES: Obadiah Stane, Aldrich Killian, Thanos
SKILLS: Business sense, combat skills
GEAR: Rescue armor

SET: 76190 Iron Man: Iron Monger Mayhem
YEAR: 2021

TO THE RESCUE
Taking a page from Iron Man's book, Pepper wears her own armored suit with thrusters and wings when she joins the Avengers in her Rescue armor.

HELEN CHO

GENIUS GENETICIST

DR. HELEN CHO MAKES her first and only appearance in LEGO® form in the Avengers Tower set (76269). She joins the Avengers at their base to heal Hawkeye using her knowledge of cell biology. Later, Ultron forces Cho to create a new body for him—but her minifigure courageously disrupts his plan!

Neat bun hair style is the same shape as Aunt May's

An alternate face shows scared expression

Sterile clothing for lab work

SUPER STATS

LIKES: Science, genetics
DISLIKES: Parties, unless Thor is going
FRIENDS: Tony Stark, Maria Hill
FOES: Ultron
SKILLS: Brilliant mind, quick-thinking
GEAR: Medical supplies, Regeneration Cradle

SET: 76269 Avengers Tower
YEAR: 2023

BODY BUILDING
Cho's pioneering medical invention, the Regeneration Cradle, builds a body from organic tissue and vibranium. The body was supposed to be for Ultron but ended up becoming Vision instead!

THE WATCHER

OBSERVER OF ALL REALITIES

Eyes glow white

Fabric collar is unique to this minifigure

One-sleeved blue cape

THE WATCHER is a being who does as his name suggests... watches. The Watcher tries his best to uphold The Watcher oath not to interfere. But when an alternate version of Ultron threatens The Multiverse, The Watcher gathers a team known as the Guardians of The Multiverse.

SUPER STATS

LIKES: The Multiverse
DISLIKES: Having to do anything but watch
FRIENDS: Doctor Strange, The Guardians of The Multiverse
FOES: Ultron
SKILLS: Omniscience, energy blasts
GEAR: Golden armor

SET: 76194 Tony Stark's Sakaarian Iron Man
YEAR: 2021

WATCH THIS!
The Watcher appears to both Tony Stark and Gamora, but it is Gamora he wants to recruit as a Guardian of The Multiverse.

VALKYRIE

KING OF NEW ASGARD

THE LAST SURVIVING member of Asgard's famous female Valkyrie warriors, this hero—who has taken on the name Valkyrie for herself—is the new king of the Asgardians. Her minifigure rules with compassion and honor and is brave enough to stand up to anyone—including the former king, her friend Thor!

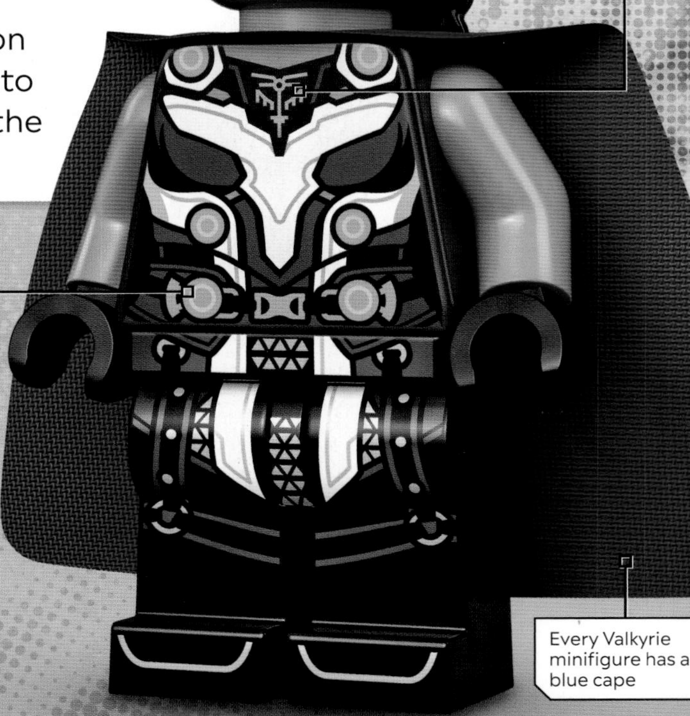

Valkyrie emblem etched on chestplate

Armor in traditional Asgardian design

Every Valkyrie minifigure has a blue cape

SUPER STATS

LIKES: Being on the battlefield
DISLIKES: Meetings, family squabbles
FRIENDS: Thor, The Mighty Thor
FOES: Hela, Thanos, Gorr
SKILLS: Combat, leadership, Asgardian powers
GEAR: Zeus's Thunderbolt

SET: 76208 The Goat Boat
YEAR: 2022

FLYING HORSE
Valkyrie rides her trusty steed, Warsong, through an interdimensional portal to take part in the battle against Thanos.

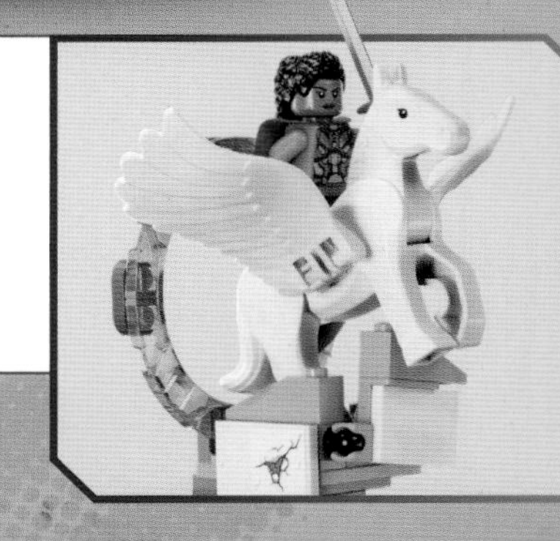

MIEK AND KORG

NEW ASGARDIANS

TWO ALIENS, MIEK AND KORG, meet Thor on the planet Sakaar and end up joining him on his adventures. Miek is a small, insectoid Sakaaran whose three-piece LEGO form sits in an exoskeleton with long arm blades. Korg's minifigure is a soft-spoken Kronan who looks like a living pile of rocks. Both are brave warriors and loyal to their new Asgardian home.

DID YOU KNOW?

Miek goes on to get a job as King Valkyrie's assistant—and is surprisingly efficient at local admin work!

WINTER SOLDIER

VILLAIN TURNED HERO

LIKE HIS OLDEST FRIEND, Captain America, James "Bucky" Barnes was injected with Super Soldier Serum. Unlike Cap, Bucky was forced to work for Hydra, though he escaped and eventually joined the Avengers. Now, Bucky's minifigure teams up with the new Cap Sam Wilson and works to make amends for his past.

Face printed with stubble and a confident smile

Vibranium arm is a gift from Shuri

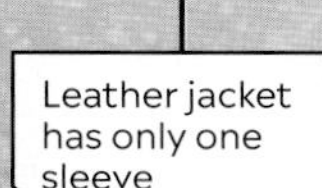

SUPER STATS

LIKES: Cake, staring
DISLIKES: When Sam Wilson talks too much, wizards
FRIENDS: Captain America, Falcon, Sharon Carter
FOES: Hydra, Thanos
SKILLS: Super Soldier abilities, combat expertise
GEAR: Vibranium arm

SET: 71031-13 LEGO Minifigures—Marvel Studios Series
YEAR: 2021

BAD BUCKY

An earlier Bucky minifigure has longer hair and a big scowl. It comes in set 76047, where the brainwashed Bucky faces off against Black Panther.

SHARON CARTER

FORMER AGENT

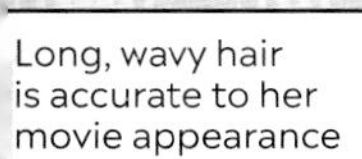

Casual black belted jacket

KNOWN AS AGENT 13, Sharon Carter works undercover for S.H.I.E.L.D. Her minifigure appears in just one set, where she helps Captain America save Bucky Barnes. After this, Carter goes on the run for years before reuniting with Bucky. But once again, she hides the truth about the nature of her work...

SUPER STATS

LIKES: Power, money
DISLIKES: People knowing that her aunt was Peggy Carter
FRIENDS: Captain America, Bucky Barnes
FOES: Brock Rumlow, Karli Morgenthau
SKILLS: Master spy, tactical mind, expert markswoman, undercover experience
GEAR: None

SET: 76051 Super Hero Airport Battle
YEAR: 2016

TOP OF THE CLASS
Carter is a S.H.I.E.L.D. agent with years of combat and espionage experience. Her most useful skill, though, is how she makes up her own mind about who to trust.

NICK FURY

SUPER SPYMASTER

FORMER DIRECTOR OF S.H.I.E.L.D. and founder of the Avengers, Nick Fury is one of the most influential people in the universe. His minifigure is instantly recognizable, thanks to its eye patch and signature black trenchcoat. Though Fury doesn't pop up in too many sets, you can be sure he's always working to save the world from the shadows.

Eye patch doesn't fully hide scars

S.H.I.E.L.D. insignia stitched on coat

Leather trenchcoat worn over dark clothing

SUPER STATS

LIKES: Law and order, most of the time
DISLIKES: When his toast isn't cut the way he likes it
FRIENDS: The Avengers, Phil Coulson, Maria Hill
FOES: Hydra, Aldrich Killian, Thanos
SKILLS: Espionage, tactical mind, years of experience
GEAR: None

SET: 76153 Avengers Helicarrier
YEAR: 2020

YOUNG NICK
The Avengers find it hard to believe, but Fury was young once. His more youthful minifigure wears a gun holster and has two working eyes.

MARIA HILL

SUPER AGENT

Alert expression—her alternate face is angry

Simple black leather jacket

Combat pants allow ease of movement

DID YOU KNOW?

Maria Hill and Bruce Banner both work with the Avengers, but they have more in common than you might think. Their minifigures have the same torso piece!

MARIA HILL'S minifigure is the only one Fury trusts completely. Whether she's working with S.H.I.E.L.D., the Avengers, or undercover aliens, Hill is always loyal, honest, and strong-minded.

SUPER STATS

LIKES: Teasing Captain America, porcupines
DISLIKES: Tight helmets, traitors
FRIENDS: Nick Fury, Captain America, Pepper Potts
FOES: Loki, Hydra
SKILLS: Combat, espionage, brilliant tactical mind
GEAR: None

SET: 40343 Spider-Man and the Museum Break-In
YEAR: 2019

HARD WORKER
Hill has appeared in only two LEGO sets. In one she's on board the S.H.I.E.L.D. Helicarrier, and in another she joins Spider-Man to protect precious objects during a museum break-in.

AGENT COULSON

SHIELD OF HUMANITY

S.H.I.E.L.D. AGENT PHIL COULSON has lived, died, and been brought back to life in his efforts to protect Earth from its deadliest threats. Coulson's minifigure wears a standard-issue suit and tie and has a headset printed on his face because he is always busy saving the world.

S.H.I.E.L.D. ID badge worn proudly

Formal work attire is Coulson's favorite outfit

SUPER STATS

LIKES: Vintage Super Hero cards
DISLIKES: People who are impolite
FRIENDS: Nick Fury, Tony Stark, Pepper Potts
FOES: Loki, Hydra, John Garrett
SKILLS: Friendliness, analytical mind, intelligence
GEAR: Cybernetic hand

SET: 76077 Iron Man: Detroit Steel Strikes
YEAR: 2017

COOL COULSON

It's not all work and no play for Coulson. His minifigure drives a cool flying car with foldable wheels, and his alternate face wears even cooler shades!

S.H.I.E.L.D. AGENT

GOVERNMENT FOOT SOLDIER

Aviator helmet for flight missions

Classic S.H.I.E.L.D. uniform with insignia

Chunky utility belt holds ammunition

S.H.I.E.L.D. AGENTS carry out missions to protect their country. Well trained and fully equipped with S.H.I.E.L.D. weapons, this agent's minifigure grins proudly in his blue uniform, ready to obey the orders of his director.

SUPER STATS

LIKES: Completing missions, praise
DISLIKES: People forgetting his name
FRIENDS: Maria Hill, Phil Coulson
FOES: Carnage, Hydra
SKILLS: Combat and weapons training
GEAR: S.H.I.E.L.D. equipment

SET: 76036 Carnage's SHIELD Sky Attack
YEAR: 2015

TIME TO PANIC
This minifigure isn't always so happy. His head piece has a worried expression, too—perfect for when his jet flyer is attacked by Carnage!

WOLVERINE

MUTANT HERO

WOLVERINE IS A MUTANT with unusual powers that include rapid healing and huge claws that extend from his hands. He joins a team of fellow mutants—the X-Men—and together they fight various vile villains. Wolverine has appeared in LEGO form several times, but this 2024 minifigure is one of the first to be based on the Marvel Animation series *X-Men '97*.

Unique helmet mold created for Wolverine

Adamantium claws can retract into hands

SUPER STATS

LIKES: Training new recruits
DISLIKES: Metal detectors
FRIENDS: X-Men
FOES: Mister Sinister, the Sentinels
SKILLS: Rapid healing, animal characteristics
GEAR: Retractible adamantium claws

SET: 76281 X-Men Jet
YEAR: 2024

WAY OF THE WOLVERINE
Using his animal senses and abilities, Wolverine approaches his foes stealthily. Then he pounces on them in a surprise attack!

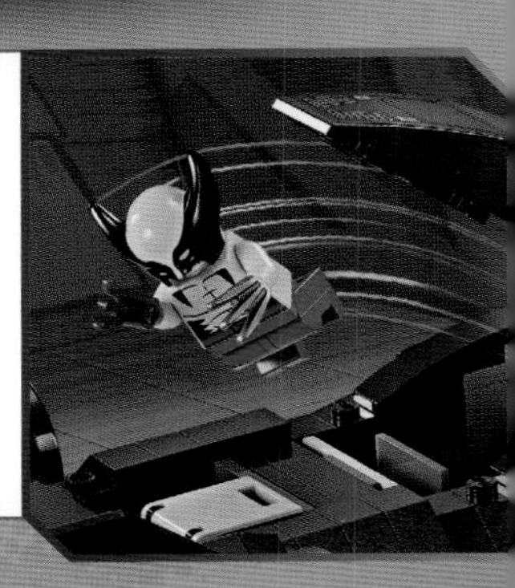

X-MEN

MUTANT TEAM

THE X-MEN TEAM is made up of individuals with a range of mutant powers. While they are not always accepted by others, they band together to fight injustice.

Ruby-quartz visor controls energy beams

X-Men emblem on chest strap clasp

CYCLOPS

Cyclops can shoot energy beams from his eyes. His minifigure wears his iconic red visor and—unlike his 2014 version—has a full head of hair on show above his face mask.

ROGUE

This 2024 minifigure is Rogue's first appearance in LEGO form. She can absorb other people's powers with a single touch. That's why her minifigure wears gloves all the time!

MAGNETO

He used to be an enemy of the X-Men, but Magneto now hopes to lead them. Three earlier Magneto minifigures wore his red-and-purple comic-book costume. This new design sees him wearing a pink suit and a friendly smile.

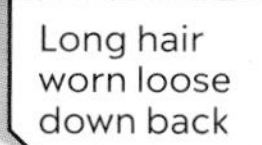

Front of cape printed on torso

White streak across front of hair

Leather jacket worn over her suit

YELENA BELOVA

REFORMED ASSASSIN

SPEEDING SISTERS

Like her "sister," Black Widow, Yelena is fearless when riding a motorcycle. This skill comes in handy when they are chased by the assassin Taskmaster.

SUPER STATS

LIKES: Dogs, vests with pockets
DISLIKES: Posers
FRIENDS: Black Widow, Red Guardian
FOES: Taskmaster
SKILLS: Combat, espionage, pilot
GEAR: None

SET: 76162 Black Widow's Helicopter Chase
YEAR: 2020

Fleece-lined pilot gloves

Combat vest with lots of useful pockets

White suit provides camouflage in snowy landscapes

DID YOU KNOW?

During their mission, Yelena and Black Widow reunite with their adoptive parents: their father, Red Guardian, and their mother, Melina.

YELENA GREW UP AS Black Widow's adoptive sister—before training to become a Black Widow assassin. Twenty years later, her minifigure teams up with Black Widow as they try to take down the entire Widow training operation that coaches assassins.

RED GUARDIAN

SOVIET PATRIOT

RED GUARDIAN'S minifigure is just as elusive as the character himself, who lived as a Soviet spy in America for three years. Appearing in just one set—a Comic-Con exclusive—Red Guardian wears his suit with pride, honored to be his country's answer to Captain America!

DID YOU KNOW?

Set 77905 was intended to be a giveaway at Comic-Con in 2020. But the event was canceled due to the COVID-19 pandemic, so the sets were sold online instead.

Beard and stubble printed on face

Utility belt print continues around back of torso

Red Guardian suit is a little too tight

SUPER STATS

LIKES: Arm wrestling, making a scene
DISLIKES: Pigs, prison
FRIENDS: Black Widow, Yelena
FOES: Taskmaster
SKILLS: Super-strength
GEAR: Shield

SET: 77905 Taskmaster's Ambush
YEAR: 2020

RED RIVALRY
Red Guardian often compares himself to Captain America. His minifigure even wields his own shield, like Cap. It's decorated with a red star.

NEBULA

CYBORG GUARDIAN

BROUGHT UP AS A daughter of Thanos, Nebula's cyborg minifigure fights against Earth's heroes—including her adoptive sister, Gamora—in three LEGO sets. But eventually Nebula realizes the truth: that Thanos is evil. So she dons a white Time Suit to help the Avengers defeat him once and for all.

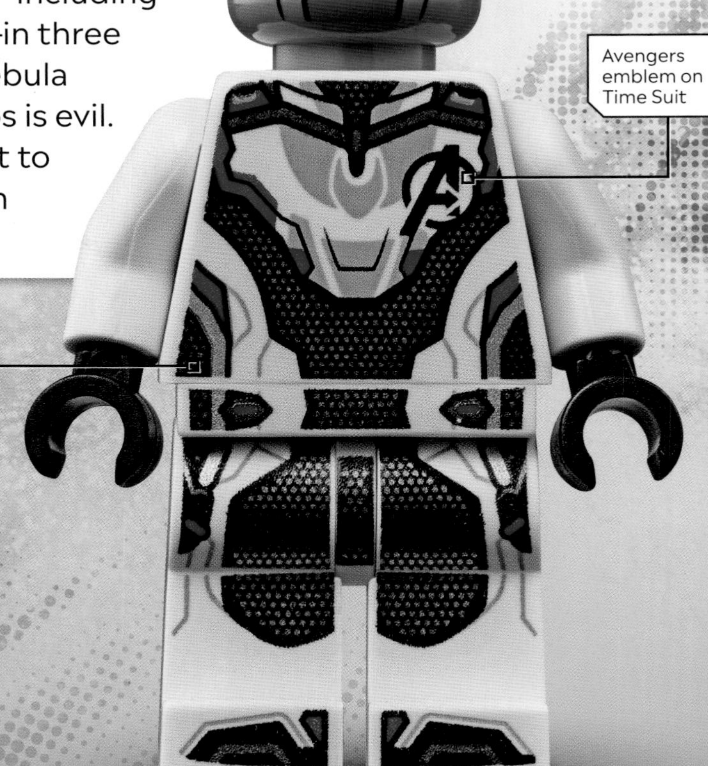

Cybernetic implant can project Nebula's memories

Avengers emblem on Time Suit

Nanotech Time Suit designed to travel through Quantum Realm

SUPER STATS

LIKES: Having a sister
DISLIKES: Having a sister
FRIENDS: Gamora, Guardians of the Galaxy
FOES: Thanos
SKILLS: Super-strength, durability, healing ability
GEAR: Cybernetic body

SET: 76131 Avengers Compound Battle
YEAR: 2019

TEAM GUARDIAN
After many ups, downs, double-crosses, and time-travel tricks, Nebula becomes an official member of the Guardians, as seen by her minifigure's new choice of suit.

YONDU UDONTA

FORMER RAVAGERS LEADER

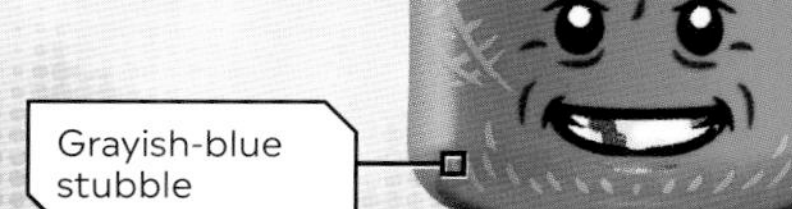

UNTIL HE WAS OVERTHROWN, the alien Yondu was leader of the Ravagers criminal clan. He is tough and ruthless—and responsible for kidnapping the young Star-Lord. Yet in the end, he shows where his loyalties lie when he sacrifices himself to save Star-Lord's life.

SUPER STATS

LIKES: Trinkets, whistling
DISLIKES: New technology
FRIENDS: Star-Lord, Kraglin
FOES: Taserface
SKILLS: Thievery, tactical mind
GEAR: Yaka Arrow

SET: 76080 Ayesha's Revenge
YEAR: 2017

QUICK GETAWAY
Having been a space pirate for many years, Yondu has had a lot of practice at getting out of scrapes!

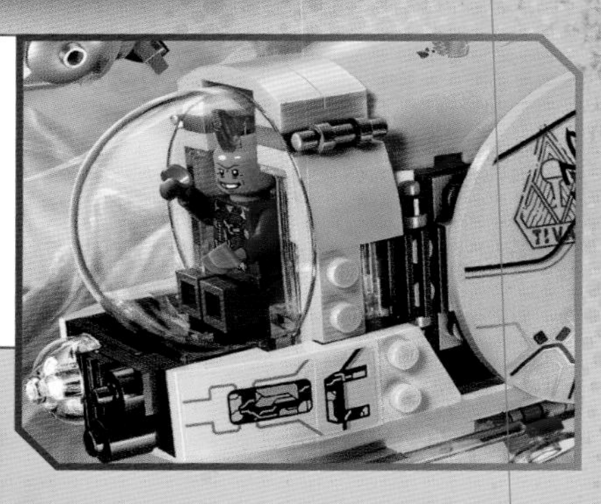

DID YOU KNOW?

Every part of Yondu's minifigure is unique, making him a firm fan favorite.

OKOYE

DORA MILAJE LEADER

THE HEAD OF WAKANDA'S special forces, the Dora Milaje, Okoye is a brave warrior. Her minifigure acts as a personal bodyguard to Black Panther and also carries out elite missions for her kingdom.

SUPER STATS

LIKES: A cup of coffee, rhinos
DISLIKES: Restrictive jackets, wigs
FRIENDS: Black Panther, the Avengers
FOES: Ulysses Klaue, Thanos, Namor
SKILLS: Combat, leadership
GEAR: Vibranium spear

SET: 76247 The Hulkbuster: The Battle of Wakanda
YEAR: 2023

DID YOU KNOW?

Okoye's minifigure comes with her trusty spear in every set except for one. In the 2023 Advent Calendar, she has an ice hockey stick instead!

FEARLESS
In battle, Okoye relies on her combat training, acrobatic strength, and vibranium spear—much to the dismay of her enemies!

NAKIA

WAKANDAN SPY

DID YOU KNOW?

Nakia's unique head piece includes white facial markings, traditionally painted on for Wakandan ceremonies.

NAKIA BUILT UP a solid reputation as Wakanda's top spy, but after the death of T'Challa, she ran away. Nakia returns after Shuri is captured by Namor, king of Talokan. Her minifigure wears a submersible suit to rescue her old pal from Namor's underwater kingdom.

Bioluminescent trim suited for underwater missions

Underwater suit in the colors of Nakia's tribe, the River Tribe

SUPER STATS

LIKES: Helping other people
DISLIKES: Vengeance
FRIENDS: Black Panther, Okoye
FOES: Ulysses Klaue, Erik Killmonger, Namor
SKILLS: Espionage, disguise, combat
GEAR: Ring blades

SET: 76211 Shuri's Sunbird
YEAR: 2022

DORA MILAJE DAYS
Nakia's first minifigure wears red Dora Milaje armor and wields her signature ring blades.

M'BAKU

JABARI TRIBE LEADER

Unique head piece with flecked beard

M'BAKU LEADS THE Jabari tribe of Wakanda. At first, his minifigure opposes T'Challa as Black Panther, but he later becomes a loyal ally. Though M'Baku helps defeat Erik Killmonger, he makes his first LEGO appearance during the battle against Namor.

DID YOU KNOW?

M'Baku once challenged T'Challa for the throne of Wakanda. When Shuri later decides to travel the world, M'Baku challenges for the throne again.

Wooden armor painted with the face of the Jabari's gorilla god

Belt made from natural fibers

SUPER STATS

LIKES: Pretending to be more scary than he is
DISLIKES: People who don't honor tradition
FRIENDS: T'Challa, Black Panther, Queen Ramonda
FOES: Erik Killmonger, Thanos, Namor
SKILLS: Combat, tribal knowledge
GEAR: Club

SET: 76214 Black Panther: War on the Water
YEAR: 2022

LOYAL WARRIOR
When Namor and his Talokanil army attack Wakanda, M'Baku is one of the first to stand up to them. He joins Black Panther and Okoye for the final battle.

IRONHEART

GENIUS STUDENT

IRONHEART MARK 2
The first Ironheart armor was built from scrap metal. Riri creates a sleeker, more high-tech Mark 2 vibranium suit in Shuri's lab.

SUPER STATS

LIKES: Differential equations, encryption, proving herself
DISLIKES: Getting locked out of her own laptop
FRIENDS: Queen Ramonda, Black Panther
FOES: Namor, Namora
SKILLS: Genius mind, inventing, problem-solving
GEAR: Ironheart suit

SET: 76211 Shuri's Sunbird
YEAR: 2022

SMART STUDENT RIRI WILLIAMS invents a machine that can detect precious vibranium—and suddenly finds herself in danger! When Shuri and Okoye offer to protect her, Riri's minifigure dons her latest invention, Ironheart armor (inspired by Iron Man). She shows that she can protect herself just fine!

SHANG-CHI

DEFENDER OF TA LO

SHANG-CHI TRIES TO LIVE a normal life, but it's hard when your father is a dangerous crime lord! Together with his friend Katy and sister Xialing, Shang-Chi travels to the world of Ta Lo. Here, his minifigure dons dragon-scale armor and prepares to battle against a monster known as the Dweller-in-Darkness.

DID YOU KNOW?
In Marvel Comics, Shang-Chi becomes a member of the Avengers.

Infinite knot symbol woven into armor

Red dragon scales protect against the Dweller-in-Darkness

Armor was a gift from Shang-Chi's mother

SUPER STATS

LIKES: Karaoke, living a normal life
DISLIKES: Working for his dad, Wenwu
FRIENDS: Katy, Xialing
FOES: Wenwu, Razor Fist
SKILLS: Martial arts
GEAR: Dragon-scale armor

SET: 76176 Escape from The Ten Rings
YEAR: 2021

BATTLE READY
Shang-Chi might work as a valet right now, but he still remembers the intensive martial arts training of his youth!

KATY

SHANG-CHI'S BEST FRIEND

KATY HASN'T GOT HER LIFE figured out yet, but she likes trying new things. She is shocked when she learns about the secret past of her best friend, Shang-Chi, and is excited to follow him on a new adventure. Her minifigure is surprised to discover that she is a natural with a bow and arrow!

Traditional Ta Lo robes include dragon scales

Flowers embroidered on trim

Protective knee wrappings

SUPER STATS

LIKES: Fast cars
DISLIKES: Eating with chopsticks
FRIENDS: Shang-Chi, Xialing
FOES: Razor Fist, Wenwu
SKILLS: Driving, singing
GEAR: Bow and arrow

SET: 76176 Escape from The Ten Rings
YEAR: 2021

SPEED CHASE
Katy drives Razor Fist's car during a getaway from the criminal group The Ten Rings. Her years as a valet are finally put to good use!

XIALING

NEW LEADER OF THE TEN RINGS

Severe hairstyle reflects personality

SHANG-CHI'S SISTER WAS always left behind—first by her father, then by her brother. But Xialing taught herself to be better than everyone else. After fighting alongside her brother to protect Ta Lo, Xialing's minifigure takes over her father's organization, The Ten Rings, and gets ready to do things her way.

DID YOU KNOW?

Xialing shares her determined head piece with another Marvel hero, Captain Peggy Carter.

Black strips made from Ta Lo bamboo

Unusual white dragon scales

SUPER STATS

LIKES: Drawing, being in charge
DISLIKES: Being overlooked
FRIENDS: Shang-Chi, Katy
FOES: Wenwu, Razor Fist
SKILLS: Martial arts, leadership
GEAR: Dragon-scale armor, rope dart

SET: 76177 Battle at the Ancient Village
YEAR: 2021

TA LO HERO
During the Battle of Ta Lo, Xialing forges an unlikely connection with the mystical dragon known as the Great Protector.

THE ANCIENT ONE

FORMER SORCERER SUPREME

Mystical markings continue over back of head piece

Purple silk sash ties around her back

Gold tunic flows down over hip and leg pieces

THE ANCIENT ONE'S enigmatic minifigure introduces Doctor Strange to the Mystic Arts. Though she appears in only one LEGO set, her reputation as a powerful sorcerer is known across The Multiverse. This is enough to protect Earth from most alien threats—but not all, as Doctor Strange quickly discovers!

SUPER STATS

LIKES: Snow
DISLIKES: Arrogance
FRIENDS: Doctor Strange, Karl Mordo
FOES: Dormammu, Kaecilius
SKILLS: Mystic Arts, spell knowledge, portal casting
GEAR: Fans, sling ring

SET: 76060 Doctor Strange's Sanctum Sanctorum
YEAR: 2016

FANS OF MAGIC
The Ancient One's minifigure comes with two mystical fans. They can be used for both attack and defense during battle.

WONG

SORCERER SUPREME

WONG WAS THE STRICT librarian at Kamar-Taj, where Doctor Strange first began his study of the Mystic Arts. Now Wong's minifigure is the Sorcerer Supreme—the greatest magician in the universe. He has a powerful command of magic, can create portals with immense skill, and is an expert at teasing Doctor Strange.

SUPER STATS

LIKES: Underground fighting clubs, tuna melts
DISLIKES: People stealing his books, TV spoilers, clowns
FRIENDS: Doctor Strange, She-Hulk, America Chavez
FOES: Kaecilius, Ebony Maw, The Scarlet Witch
SKILLS: Mystic Arts, portal casting
GEAR: Sling ring

SET: 76205 Gargantos Showdown
YEAR: 2022

DREAM TEAM
Wong and Strange make a great team. Together they protect the Sanctum Sanctorum from all sorts of evil threats.

AMERICA CHAVEZ

MULTIVERSE TRAVELER

Eyes glow white when creating a portal

Pendant inscribed with initials "AC"

Pin in the shape of a LEGO head piece

AMERICA CHAVEZ CAN TRAVEL across The Multiverse at will. Her minifigure has spent years leaping from universe to universe without being fully in control of this power. Scary! After meeting Doctor Strange, she starts learning to harness her powers—but first she must escape from The Scarlet Witch, who wants to take them for herself!

SUPER STATS

LIKES: Pizza balls
DISLIKES: The idea of a spider-themed hero
FRIENDS: Doctor Strange, Wong
FOES: The Scarlet Witch
SKILLS: Creating portals, super-strength
GEAR: None

SET: 76205 Gargantos Showdown
YEAR: 2022

GARGANTOS GRAB
Chavez is no stranger to weird things. However, that doesn't mean she enjoys being chased by a many-tentacled Gargantos monster!

MARIA RAMBEAU

ACE PILOT

MARIA RAMBEAU AND CAROL DANVERS were best friends during their Air Force training, until Danvers went missing. Years later, she returns as the Super Hero Captain Marvel. Rambeau's minifigure is shocked but happy and uses her top pilot skills to help Danvers defeat the evil Kree. Rambeau goes on to found the intelligence agency S.W.O.R.D.

DID YOU KNOW?

Maria's head piece is unique, but it shares one expression with the head piece of her daughter, Monica Rambeau.

Khaki jumpsuit beneath flight vest

Breathing apparatus printed on torso

SUPER STATS

LIKES: Her grandmother's cooking, running
DISLIKES: Being called "young lady"
FRIENDS: Carol Danvers, Nick Fury, Captain Marvel
FOES: Minn-Erva, Kree
SKILLS: Piloting expertise
GEAR: None

SET: 77902 Captain Marvel and the Asis
YEAR: 2019

ELUSIVE EXCLUSIVE

Rambeau is a fighter pilot who goes by the call sign "Photon." She appears in just one set, a very rare 2019 Comic-Con exclusive, along with Captain Marvel.

MONICA RAMBEAU

SPACE INVESTIGATOR

Hair piece also worn by Ironheart

Marvels emblem displayed proudly

Suit offsets the pressures of space

FOLLOWING IN HER MOTHER'S FOOTSTEPS, Captain Monica Rambeau joins S.W.O.R.D., where she gains super-powers on one of her missions! Later, she teams up with Captain Marvel and Ms. Marvel against a new Kree threat. Two similar Monica minifigures have been released, one with the S.W.O.R.D. emblem visible on her white uniform and one showing the Marvels insignia.

SUPER STATS

LIKES: Photo albums, memorabilia
DISLIKES: Multiversal rifts
FRIENDS: Captain Marvel, Ms. Marvel
FOES: Kree, Agatha Harkness
SKILLS: Flight, energy manipulation
GEAR: None

SET: 76232 The Hoopty
YEAR: 2023

HOOPTY CREW
Harnessing the power of light, Rambeau and the Marvels travel on board the spaceship called Hoopty.

AJAK

ORIGINAL PRIME ETERNAL

THE FIRST PRIME ETERNAL, Ajak leads her team of immortal Eternals to Earth, where they protect humans from deadly beasts known as Deviants. Ajak's minifigure uses her healing powers and wise guidance to help the people she meets.

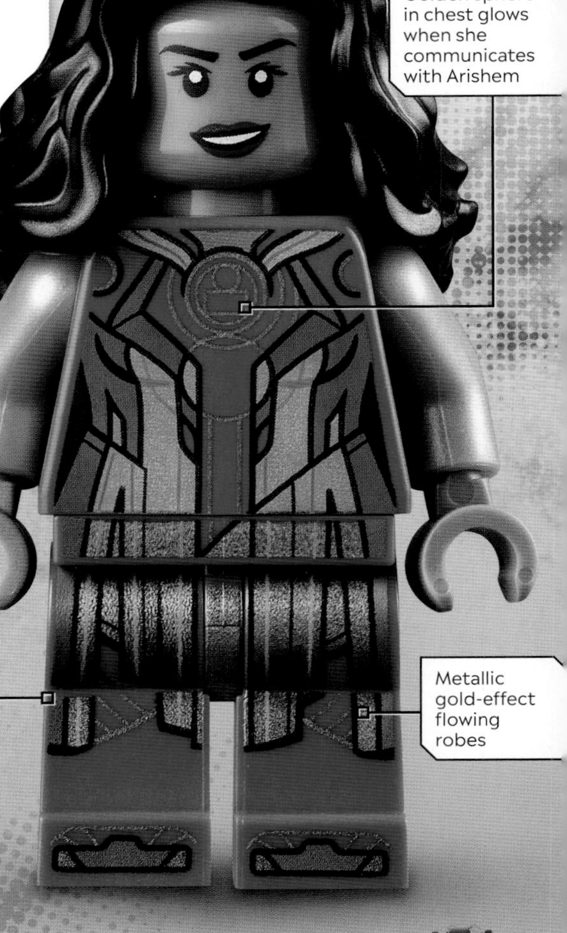

SUPER STATS

LIKES: Planet Earth
DISLIKES: Suffering
FRIENDS: Sprite, Ikarus, Sersi
FOES: Deviants
SKILLS: Healing, leadership
GEAR: Celestial armor

SET: 76155 In Arishem's Shadow
YEAR: 2021

DID YOU KNOW?

The printing on Ajak's head piece was first used for her, though it has been used on several other minifigures since.

ARISHEM'S ORDERS

Ajak receives her orders from Arishem, an infinitely powerful Celestial. She communicates with him telepathically through a sphere in her chest.

SERSI

PRIME ETERNAL

DID YOU KNOW?

Each Eternal is known by a single name, but Sersi has adapted to fit in with modern life, calling herself "Sylvia."

SERSI CAN TRANSFORM objects through touch, but her minifigure is most touched by the humans she meets on Earth. After Ajak's death, Sersi becomes the Prime Eternal, though she is guided by her heart rather than the orders of Arishem.

Geometric Celestial patterns on suit

Hands can transform whatever they touch

Silver highlights on flexible green armor

SUPER STATS

LIKES: Human beings, dancing, her cellphone
DISLIKES: Violence
FRIENDS: Ikarus, Sprite
FOES: Deviants
SKILLS: Matter transformation
GEAR: Celestial armor

SET: 76155 In Arishem's Shadow
YEAR: 2021

ETERNALS HQ

The Eternals arrived on Earth around 7,000 years ago in a huge ship named the *Domo*, which they still use as their base in times of trouble.

IKARIS

ETERNAL WITH A SECRET

SUPER POWERS
Ikaris can fly and shoot energy lasers from his eyes. Both are very useful powers when battling dangerous Deviants!

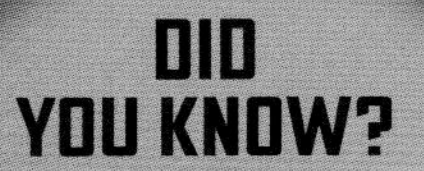

This head piece has a heroic look. It's not surprising, then, that it's also used for Captain America and Hawkeye minifigures!

SUPER STATS

LIKES: Being the favorite
DISLIKES: Going against Arishem
FRIENDS: It's complicated
FOES: Deviants
SKILLS: Flight, eye lasers
GEAR: Celestial armor

SET: 76145 Eternals' Aerial Assault
YEAR: 2021

IKARIS'S FORMIDABLE POWERS make him perhaps the mightiest of the Eternals. Many of his friends were surprised when Sersi became the new Prime Eternal instead of him. But his minifigure will be pleased to know that it does appear in more LEGO sets than any other Eternal—three!

SPRITE

ETERNAL CHILD

DID YOU KNOW?

Sprite is the only Eternal with no leg printing. In fact, short leg pieces are rarely printed!

Same hair piece style as Druig

SPRITE IS JUST AS OLD and immortal as the other Eternals, though she has the appearance of a teenager. Her minifigure comes in just one LEGO set, alongside fellow Eternal Ikaris, who she has a secret crush on!

Back printed with Sprite's cape

Sprite is the only Eternal with short legs

SUPER STATS

LIKES: Telling the truth, cellphones
DISLIKES: Hugs, video cameras
FRIENDS: Ikaris, Sersi
FOES: Deviants
SKILLS: Casting illusions, storytelling
GEAR: Celestial armor

SET: 76145 Eternals' Aerial Assault
YEAR: 2021

CONJURER OF ILLUSIONS

Sprite's illusions fool everyone, including the Deviants. They are also a good way to play pranks on her fellow Eternals!

GILGAMESH

ETERNAL PROTECTOR

FIST FIGHT
Battling a deadly Deviant, Gilgamesh's golden fists pack a powerful punch. Sadly, his powers are not enough, and Gilgamesh loses his life.

Head piece designed especially for Gilgamesh

SUPER STATS

LIKES: Vacations, witty comebacks
DISLIKES: Abandoning his friends
FRIENDS: Thena
FOES: Deviants
SKILLS: Creating armor made from cosmic energy
GEAR: Celestial armor

SET: 76154 Deviant Ambush!
YEAR: 2021

Hands punch hard enough to knock out a Deviant

Green and gold knee-length tunic

DID YOU KNOW?
Gilgamesh brews his own unique drink. He ferments it by chewing and spitting out kernels of corn!

THE STRONGEST ETERNAL, Gilgamesh's stomps and punches literally make the earth quake. But he is kind, too, caring for Thena when an illness makes her unpredictable. Gilgamesh's minifigure is also known for baking delicious pies!

THENA
ETERNAL WARRIOR

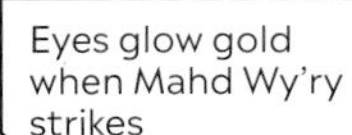

Striking high cheekbones shape Thena's face

Armored bodice protects during battle

Suit is in ancient Greek style

THENA'S IMPRESSIVE BATTLE prowess is enhanced even more by her ability to form various weapons from cosmic energy. Unfortunately, her minifigure suffers from an illness known as Mahd Wy'ry, which makes her dangerous at times.

SUPER STATS

LIKES: Memories
DISLIKES: Cowering behind walls
FRIENDS: Gilgamesh
FOES: Deviants
SKILLS: Forming cosmic energy weapons
GEAR: Celestial armor

SET: 76154 Deviant Ambush!
YEAR: 2021

CLASH WITH KRO
Furious over the loss of Gilgamesh, Thena takes on Kro, the Deviant leader. She destroys him with her golden naginata.

PHASTOS

ETERNAL INVENTOR

GIFTED PHASTOS HAS MADE countless technological and scientific contributions to humanity. His brain keeps coming up with useful ideas—but some are way ahead of his time! Despite his brilliance, his minifigure appears in only one LEGO set.

SUPER STATS

LIKES: Inventing things, his family
DISLIKES: Parties, sarcasm, plows
FRIENDS: Eternals
FOES: Deviants
SKILLS: Innovation, creating technology from cosmic energy, holographic projection
GEAR: Celestial armor

SET: 76156 Rise of the Domo
YEAR: 2021

INSIDE THE LAB
Phastos projects his ideas as holograms. He tweaks and refines them before transforming them into real objects.

KINGO

ETERNAL JOKER

Alternate face shows a quizzical expression

Loose purple tunic with gold trim

DID YOU KNOW?

The Eternals never age, so Kingo pretends that his appearances in old movies were actually his father, grandfather, great-grandfather, and great-great-grandfather!

KINGO HAS BEEN HAPPILY living as a Bollywood actor for around 100 years! When the Eternals reunite, he is pleased to see his old friends but less pleased to be drawn back into battle. Still, his minifigure's bright smile always cheers up the rest of the team.

SUPER STATS

LIKES: Making movies
DISLIKES: Conflict, flying coach
FRIENDS: Karun, Eternals
FOES: Deviants
SKILLS: Cosmic energy blasts, acting
GEAR: Celestial armor

SET: 76155 In Arishem's Shadow
YEAR: 2021

PREFERRING PEACE

Kingo's energy blasts are very effective against the Deviants. However, his minifigure would prefer to enjoy a peaceful life.

DRUIG

ETERNALLY MOODY

DRUIG'S MIND-CONTROL ABILITY gives him enormous power, but his minifigure struggles with knowing when to use it and when to allow humans to make their own mistakes. Druig always says what he thinks, even if it leads to arguments, but he has a soft spot for fellow Eternal Makkari.

Intricate Celestial patterns

Long red-and-black robes

Robe print continues over leg pieces

SUPER STATS

LIKES: Peace and quiet
DISLIKES: Hugs
FRIENDS: Makkari
FOES: Deviants
SKILLS: Mind control
GEAR: Celestial armor

SET: 76156 Rise of the Domo
YEAR: 2021

DID YOU KNOW?

If you think Druig's minifigure looks familiar, it may be because it shares a head piece—and hair piece—with Happy Hogan! (Happy's hair piece is black, though.)

HEADING HOME

Druig spent years controlling a small tribal village. When the Eternals reunite and ask Druig for his help, he sets his village free and heads back to the *Domo*.

MAKKARI

ETERNALLY SWIFT

Long hair is braided and coiled around her head

DID YOU KNOW?

Makkari is the first minifigure with this braided hair piece. It is a new mold that was created especially for her.

MAKKARI IS DEAF, but she hears everything, thanks to her ability to sense even the slightest vibrations. She can also run faster than the speed of sound, crossing continents and oceans in seconds. Her minifigure fights the Deviants in two sets.

Suit is a mix of metal and fabric

Red-and-gray protective boots

SUPER STATS

LIKES: Collecting treasures
DISLIKES: Thieves, boredom
FRIENDS: Druig
FOES: Deviants
SKILLS: Super-speed, can sense vibrations
GEAR: Celestial armor

SET: 76154 Deviant Ambush!
YEAR: 2021

HOME SWEET HOME

Makkari has lived on board the *Domo* for years. It is where she keeps her collection of antiques—and her favorite teacup!

Senior Editors Laura Gilbert, Laura Palosuo
Designers David McDonald,
Isabelle Merry, Samantha Richiardi
US Senior Editor Jennette ElNaggar
Senior Production Editor Jennifer Murray
Senior Production Controller Lloyd Robertson
Managing Editor Paula Regan
Managing Art Editor Jo Connor
Managing Director Mark Searle
Jacket Designer James McKeag

DK would like to thank Ashley Blais, Randi K. Sørensen, Heidi K. Jensen, Martin Leighton Lindhardt, and Adam Corbally, Justin Ramsden, Mark Tranter, and the rest of the LEGO Marvel Super Heroes Design team at the LEGO Group; Chelsea Alon at Disney Publishing; and Lauren Bisom at Marvel Comics. Special thanks to Kevin Feige, Louis D'Esposito, Brad Winderbaum, Kristy Amornkul, Sarah Beers, Capri Ciulla, Jacqueline Ryan-Rudolph, Erika Denton, Nigel Goodwin, Jennifer Giandalone, Jennifer Wojnar, and Jeff Willis at Marvel Studios. DK also thanks Julia March for proofreading this book.

First American Edition, 2024
Published in the United States by DK Publishing,
a division of Penguin Random House LLC
1745 Broadway, 20th Floor, New York, NY 10019

24 25 26 27 28 10 9 8 7 6 5 4 3 2 1
001–339775–Oct/24

Manufactured by Dorling Kindersley, One Embassy Gardens, 8 Viaduct Gardens, London SW11 7BW, under license from the LEGO Group.

Published in Great Britain by Dorling Kindersley Limited

A catalog record for this book
is available from the Library of Congress.
ISBN 978-0-5938-4392-5
LIB ISBN 978-0-5938-4712-1

Printed and bound in China

www.dk.com